Rhododendrons in India: Floral & Foliar Splendour of the Himalayan Flora

Rhododendrons in India: Floral & Foliar Splendour of the Himalayan Flora

A.R.K. Sastry
Principal Investigator*

P.K. Hajra

**Project Supported by*
Department of Science & Technology, Government of India,
Technology Bhavan, New Meharauli Road,
New Delhi, under its USERS' Scheme.

BSP **BS Publications**
A unit of **BSP Books Pvt. Ltd.**
4-4-309/316, Giriraj Lane, Sultan Bazar,
Hyderabad - 500 095
Phone : 040 - 23445605, 23445688

Cover Photos

Species Names:

Top:	**Left to Right:**	*R. keysii*, *R. cinnabarinum* ssp. *cinnabarinum*.
Below:	**Left to Right :**	*R. dalhousiae* var. *rhabdotum*, *R. maddenii* ssp. *maddenii*.
Below:	**Right Bottom :**	*R. cinnabarinum* ssp. *xanthocodon*

Published by :

BSP BS Publications
A unit of **BSP Books Pvt. Ltd.**
4-4-309/316, Giriraj Lane, Sultan Bazar,
Hyderabad - 500 095
Phone : 040 - 23445688, 23445605
e-mail : info@bspbooks.net

ISBN : 978-93-52300-42-6 (HB)

Addala Rama Murty & Suramma (Parents)

A.L. Narasimha Murty & Venkatalakshmi

In Memoriam

To my parents; my eldest brother and sister-in-law for their care and affectionate bringing-up from my very childhood.

A.R.K. Sastry

Dedication

This book is dedicated to Prof. Dr. Rolla Seshagiri Rao, former Director-in charge, Botanical Survey of India, Kolkata, for his pioneering post-independent botanical explorations to Mt. Cho Oyu, Sikkim, Arunachal Pradesh and other N.E. Indian Hilly states which enriched our knowledge on the flora of the region and for his important and valuable contributions on the vegetation and taxonomy of Indian flora

and

to the people of Sikkim, Arunachal Pradesh and N.E. India for preserving and protecting the rich biodiversity for centuries.

- Authors

PROF. ROLLA SESHAGIRI RAO
M.Sc., Ph.D., F.L.S.(London), F.B.S., F.A.A.Sc., F.N.E.S.A., F.I.A.T.
Formerly-Director, Botanical Survey of India; Professor & Head
Botany Dept., Andhra University.; Director, DST., BSI., CSIR & UGC. Flora Projects
President, Society for Conservation & Management of Natural Resources
Consultant: Indian Flora, Medicinal Plants, Environment & Biodiversity
Member: INTACH, AP State Chapter Consultative Committee; Governing Council, New Delhi
Ramananda Tirtha Institute for Medicinal & Aromatic Plants Committee

Foreword

Inspite of having a good network of R & D establishments, National and International Policy frame work, adequate natural and human resource wealth, advancements in Biotechnology and Information Technology, India's achievements in biodiversity and related areas of research did not receive much attention. Lack of co-ordination and integrated efforts, reluctance to accept and support basic science disciplines like Systematics, lack of collective vision and common agenda for tackling complex research and development, are seen as some of the inherent weaknesses of our R & D systems particularly with reference to Systematic Biology research. It is therefore, necessary to examine the problems of Systematic Biology research and suggest to evolve a national mechanism to coordinate all on-going and future researches in Systematics and Taxonomy and to integrate it to meet the national goals of inventory, conservation pospecting and sustainable use of biodiversity resources of our country.

With such background, Society for Conservation and Management of Natural Resources (Regd.) was started in 1992 at Hyderabad by a group of Scientists and enthusiasts in scientific studies. The society holds an excellent track record of undertaking research projects on conservation of rich Biodiversity in India and has brought out important publications covering plant groups (District Floras, lower groups like Fungi, Mangroves and Medicinal plants of Andhra Pradesh), animal and fisheries groups like Reptiles, Molluscan groups, Meiofauna of sandy beaches, geographical aspects of Andhra Pradesh etc. The present publication on Indian Rhododendrons is another such important study successfully undertaken and completed by A.R.K. Sastry, the Principal Investigator for this project sponsored by the Department of Science & Technology, Govt. of India, New Delhi under 'Retired Scientists USERS' Programme, along with the junior author P.K. Hajra who have worked with me in the Botanical Survey of India.

I am how delighted to write a Foreword to the publication of 'Rhododendrons in India: Floral and Foliar Splendour of the Himalayan Flora', a subject very dear to me and extensively studied during my plant explorations along the Himalayas of India.

The Indian Rhododendrons constitute an important element in the flora of the entire Himalayan ranges and in the hilly states of N.E India and have attracted the attention of great botanists like J.D. Hooker and others who introduced them into the Botanical gardens in England since the 19th century. Indian Rhododendrons have contributed a great deal to the horiculture in the West and are widely grown in temperate countries for their very attractive flowers.

The information presented in this book will no doubt be of interest to botanists, horticulturists, teachers, students, and nature lovers and will aid in the correct taxonomic identification of the Rhododendrons growing in India and their conservation. Of late, this beautiful flowering genus of the Himalayas is being very much disturbed and cut for fuel along the Himalayan States of India and more particularly in Arunachal Pradesh. Better protection measures are to be adopted urgently to conserve this Himalayan beauty by the concerned departments for safe-guarding the clourful environment of the Himalayas.

I congratulate the authors for this valuable contribution which I am sure will fill the lacuna on the identification of the species in India and help promoting conservation of this very important constituent in the Himalayan flora.

Hyderabad

Date: 19th June, 2009

R.S. Rao

Preface

With over a 1000 species world wide rhododendrons are acclaimed in horticulture and have received a great deal of attention for taxonomical studies since Linnean times. Much work has been done on their taxonomy and horticultural aspects by various institutions and botanists in the Western world and many important publications have been brought out.

The Himalayan rhododendrons have attracted the attention of many botanists since J.D. Hooker's botanical explorations in 1840s in Sikkim and Khasia hills (in Meghalaya). With very attractive, usually bell-shaped large flowers of varying colours in dense floral bunches and equally elegant foliage, Indian Rhododendron species have contributed richly in horticulture in the West.

Despite several botanical explorations and publications on the flora of the Himalayan region in India, no attempt has been made to bring out a consolidated account of the Rhododendron species that occur in the Indian region detailing their taxonomy, habit and distribution, ecology, conservation status etc. Of the species that are distributed in the Eastern Himalaya, only the species that occur in Sikkim have been dealt in their publications by J.D. Hooker, C.B. Clarke and U.C. Pradhan & S.T. Lachungpa. The Rhododendron species nearing to around 70 in number that inhabit the mountains in Arunachal Pradesh and other hilly states in N.E. India have not been worked out taxonomically and thereby leaving a wide gap in our knowledge. In order to fill this lacuna the project has been taken up by the Senior author and Principal Investigator in the year 2004 with the financial assistance from the Department of Science & Technology, Govt. of India, New Delhi under their scheme: Utilisation of Scientific Expertise of Retired Scientists (USERS). The senior author while on a private visit to U.K in the same year made special efforts to examine the Indian specimens of rhododendrons deposited in the herbaria of Kew and Edinburgh Royal Botanic Gardens, U.K and gathered important relevant literature and photographed some of the Type specimens for authentic studies on the species in India. Besides, specimens in the Botanical Survey of India herbaria at Kolkata, Shillong, Dehradun and FRI herbarium were examined and notes from the published literature combined with field observations made during botanical explorations in Sikkim, Arunachal Pradesh and parts of Meghalaya have also been taken into consideration in writing this book.

In this book, a general introductory account on the genus, its classification, taxonomic keys to the different subgenera, sections, subsections, species and infra specific categories, latest botanical names with citations and references, scientific descriptions, notes on their distribution and habitat etc. have been given. 26 maps showing distribution of all the taxa and line drawings for 19 species together with colour photographs for several species are included to make this publication scientific and user friendly.

It is hoped that this book will help in the identification of Indian Rhododendrons by students of Botany, teachers, nature lovers, foresters and conservationists and will promote their conservation in natural habitats and in furtherance of scientific work.

- Authors

Suggestions are welcome by the author and can be contacted at the following address:

Address:

A.R.K. Sastry

S-8, Yechuri Towers,

Sesha Sai Street, Rama Rao Peta,

KAKINADA - 533 004, A.P. India.

Phone : (Home) 0884 - 2368711

Mobile : 09848436805

e-mail : arksastry@yahoo.com

Acknowledgements

My sincere thanks are placed on record to the Department of Science & Technology (DST), Govt. of India, New Delhi for providing financial support under its Utilisation of Scientific Expertise of Retired Scientists (USERS) Scheme enabling to complete writing of this book.

I sincerely thank Dr. Rolla Seshagiri Rao, President, and the Secretary of the Society for Conservation and Management of Natural Resources (SCMNR), Hyderabad, and Prof. H.Y. Mohan Ram, F.N.A., former Prof. & Head, Department of Botany, Delhi University for their help, encouragement and suggestions.

Prof. Simon J. Ownes, Keeper and Dr. Jonathan, Kew Herbarium; Dr. David J. Harris, Keeper and Dr. Adele Smith, Assistant Curator, Edinburgh Herbarium gave permission to consult their Herbaria and libraries which is appreciated and thankfully acknowledged. I also express my sincere thanks to Prof. D.F. Chamberlain, Dr. George Argent and Dr. Henna Aitkins for their valuable time, discussion on the subject and suggestions while working in the Edinburgh Herbarium. My sincere thanks are also due to my friends Mr. Peter A. Cox, Mrs. Patricia Cox and Mr. Kenneth E. Cox for their suggestions, warm hospitality during my visit to their home and presenting a copy of their book.

My sincere thanks are due to Dr. M. Sanjappa, Director, BSI for help in many ways; Dr. P. Lakshmi Narasimhan, former Indian Botanical Liason Officer (IBLO) of BSI, Kew Herbarium for help and providing photocopies of some important literature from the library; Drs. P. Venu, G.V.S. Murthy, H.J. Chowdhury, J.R. Sharma, S.K. Srivatsava (IBLO, BSI at Kew), P.G. Diwakar, Jayashree Bhattacharya, V. Sampath Kumar of BSI and Dr. (Miss) Veena Chandra, FRI Herbarium, Dehra Dun for their help. Mr. Sudipto Chatterjee and Piyush Dutta formerly working with me in the WWF-India, New Delhi provided some photographs which is gratefully acknowledged. My sincere thanks are also due to Mr. Shyam Chivukula and Mrs. Padmini Chivukula (my son-in-law and daughter), Farnbourgh, Kent, U.K. for organizing my visit to Edinburgh and Glencarse in Scotland in September 2004; to my wife Kamala for her support and daughters : Uma and Madhavi and sons-in-law Prasad and Bhaskar for their good wishes.

Last but not the least I sincerely thank Mr. N. Veerabhadra Rao, Technical Assistant, M.S. Swaminathan Research Foundation, Kakinada for meticulous and neat computer type-setting of the draft manuscripts and help in various other ways and to Dr. R. Ramasubramanian, Scientist-in-change, MSSRF for help.

A.R.K. Sastry

The authors express sincere thanks to many of their former colleagues in the Botanical Survey of India, particularly Dr. M. Sanjappa, Director and the authorities of FRI Herbarium, Dehra Dun for their help.

Sincere thanks are placed on record to Mr. Nikhil Shah, Mr. Anil Shah, Mr. Naresh Davergave and his team of BS Publications, Hyderabad for their excellent co-operation and early publication.

- Authors

Contents

Photographic Credits

A.R.K. Sastry

1, 4, 5, 6, 13, 14, 15, 16, 17, 18, 20, 26, 29, 30, 31, 36, 37, 37a, 38, 39, 39, 42, 43, 46, 47, 53, 54, 56, 57, 58, 59, 60, 62, 69, 70, 70a, 73, 74, 75, 76, 78 and all the photographic of colour drawings from J.D. Hooker's Rhododendrons of Sikkim Himalaya and photographs of the Type materials of some Rhododendron species taken in Kew Herbarium.

Sudipto Chatterjee and Piyush Ghosh, WWF-India

3, 7, 8, 9, 10, 19, 21, 22, 23, 24, 25, 27, 28, 33, 34, 35, 40, 41, 44, 45, 48, 49, 50, 51, 52, 55, 61, 63, 64, 65, 66, 67, 68, 71, 72, 77

P.K. Hajra

2, 11, 12

1
Introduction

Introduction

The genus *Rhododendron*, a member of the family Ericaceae, was founded by Carl Linnaeus (*Species Plantarum*, 1753). The word *Rhododendron* is derived from two Greek words, namely *rhodon* (rose) and *dendron* (tree), possibly referring to the earlier known alpine rose. The first species of the genus described and named by Linnaeus was *R. ferrugineum*, which forms the type species for the genus.

1. Classification and Taxonomy: Some historical aspects

The classification and taxonomy of the genus has been a challenging task to many celebrated botanists since the Linnaean times when the genus was known chiefly from European and North American species. With the subsequent treatment of genera like *Azalea, Hymenanthes, Pentanthera, Rhodora,* to name a few, as subgenera of *Rhododendron*, the number of species have increased. The sustained efforts of great botanical explorers like Sir J.D. Hooker in the Sikkim Himalayas; George Forrest, E.H.M. Cox, F. Kingdon-Ward, Farrer, Ludlow, Sheriff & Taylor and such others covering extensive areas particularly in the Himalayas (from Kashmir to Arunachal Pradesh), N.E. Hilly states in India; in China, Burma and many parts in N. America and S.E. Asia have brought forth a wealth of rhododendron collections during the 19th and 20th centuries. The overwhelming collection of specimens involving hundreds of species from extensive geographic areas exhibiting morphological variations and even natural hybrids posed a problem in their identification and classification. The Royal Botanic Gardens at Kew and Edinburgh, became important centres of taxonomic studies on the genus. At Edinburgh under the very able directorship of Sir Issac Bayley Balfour during the period 1916-1922, great efforts on the classification and taxonomy of the genus resulted in a system of classification. Happily, the Herbaria and the Gardens at Kew and Edinburgh became world class repositories of almost all the collections from the

Himalayas, China and S.E. Asia with renowned scientists continuing their researches on the taxonomic revision even to this day. Later taxonomists like J. Cowan, J. Hutchinson, J.B. Stevenson, Rehder, Kingdon-Ward, R.H. Davidian, J. Cullen, H. Sleumer, W.R. & M.N. Philipsons, D.F. Chamberlain, George Argent and such others strived to evolve the presently accepted taxonomic classification.

The abundance of rhododendrons in this region prompted later explorers and horticulturists like Peter Cox, Sir Peter Hutchison and Kenneth Cox from Scotland to continue studies since mid 1960s.

Despite such continued research on the genus, differences of opinions still exist on the classification and taxonomy especially between the classical plant taxonomists and horticulturists. The history of classification of the genus is best concluded in the words of an expert on the subject Prof. J. Cullen of the Royal Botanic Garden, Edinburgh thus: "it is a history of tension between horticultural based and herbarium based classifications".

2. Global distribution of the genus

The genus with over a thousand species is one of the largest and most diverse among the flowering plants. The species chiefly occur in the N. Hemisphere and are found in the altitudinal range from 1000 m (*R. arboreum*) to snow bound areas at 5800 m (*R. nivale* ssp. *nivale*). Maximum concentration of species occurs in the areas where the Eastern Himalaya, the Xizang (Tibetan) Plateau, and the mountain ridges of Myanmar (Burma) and S.W. China meet. Some other species also occur in the Arctic region. The tropical or the Malaysian species which are mostly grouped under the section *Vireya* of the subgenus *Rhododendron* are found in the area between and including the Malay Peninsula extending up to the northern tip of Australia with their centre of distribution and greatest density and diversity on the Island of New Guinea.

The exact place of origin of the genus is much debated. N.W. Yunnan in China with more than 500 species is treated as the prime centre of diversity from where ancient members of the group are supposed to have moved in several directions, adopting themselves to various climatic conditions. Secondary centres of distribution are in southern Asia and tropical islands including New Guinea with about 225 species. Japan with 53 species, and Korea, Manchuria, Mangolia, Kamchatka with about 14 species; the

Arctic region with 4 species and North America with 35 species are considered as other centres (Leppik, 1974).

The total number of species, subspecies, varieties and natural hybrids which are now recognized as 'good species' is estimated to be close to 1025 world wide.

3. Historical perspectives of Indian rhododendrons - some mile stones

The history of the Himalayan rhododendrons began with the visit of Capt. Hardwick to the Siwalik mountain ranges in Kashmir in the year 1796 where he encountered a species of the genus. It was later described by Sir James Smith in his *Exotic Botany* in 1805 under the name *R. arboreum* meaning the 'tree rhododendron'. With the introduction of its seeds in England by Dr. N. Wallich, one of the Superintendents of the then Royal Botanic Garden, Calcutta (now Indian Botanic Garden under the Botanical Survey of India) in 1827, the interest in the Himalayan rhododendrons began.

Sir J.D. Hooker's journeys to the Sikkim Himalaya between 1848-1850 opened the doors to the rhododendrons of this area. During his travels, he collected and described 34 new species and subsequently published his treatise: *Rhododendrons of Sikkim Himalaya* in 1849 with beautiful natural size colour paintings. **(See Photos 1-18 of Hooker's plates).** He distributed the seeds collected in Sikkim to the gardens in England, Scotland and Wales, many of which are reported growing even today. Subsequently, C.B. Clarke (1882) revised the genus for J.D. Hooker's *Flora of British India* recording 43 species with another 3 species dealt as 'imperfectly known'. Since then many explorers like F. Kingdon-Ward (Eastern Himalaya, N.E. India, Tibet, China and Burma), Ludlow & Sheriff (Bhutan) and Ludlow, Sheriff and Taylor (Kashmir) under took many rhododendron collection expeditions during the period : 1925-1950. It was Kingdon-Ward's extensive and intensive explorations that revealed the hidden rhododendron treasure in the regions explored by him. His specimens and seeds were deposited in the herbaria and the gardens at Kew and Edinburgh and Royal Horticultural Society, London. Indian rhododendrons are best seen in these places.

Since the establishment of Eastern Circle (Shillong) and Northern Circle (DehraDun) of the reorganized Botanical Survey of India (B.S.I.) in the year 1954, several explorations were undertaken in the Western and Eastern Himalayan regions and other hilly ranges of N.E. India by Drs. R.S. Rao, G. Panigrahi, D.B. Deb, G.V. Subba Rao, J. Joseph and A.R.K. Sastry (present Principal Investigator), A.S.Rao, N.P. Balakrishnan (Bhutan), P.K. Hajra (co-author of this book) and a few others from the Eastern Circle of the B.S.I. at Shillong. Other botanists like Drs. M.A. Rao, T.A. Rao, N.C. Nair and U.C. Bhattacharya and others from the Northern Circle of B.S.I collected different rhododendrons from the Western Himalayan region.

Rhododendron species were listed in several publications and floristic accounts of many areas in the western and eastern Himalayas and north-east India by these explorers. U.C. Pradhan and Sonum T. Lachungpa (1990) published an account of the species in Sikkim. Mao, et al. (2001) published an enumerative paper on the species in India. Currently, the B.S.I. has taken up taxonomic studies on the genus in India.

4. Distribution in India

During the present study, the Principal Investigator (A.R.K. Sastry) has been able to record 92 species, 8 subspecies and 9 varieties aggregating to 109 taxa in the Indian region and are dealt in this book. Of these, a few species are distributed throughout the Himalays from Jammu & Kashmir to Arunachal Pradesh. The western Himalayan region from J.&K. to Uttaranchal has 6 species only, while the entire eastern Himalayan region in India comprising Sikkim, Darjeeling Gorkha Hill Council Area under West Bengal and Arunachal Pradesh is represented by 75 species. The maximum species diversity is seen in Arunachal Pradesh, especially in its eastern most part which lies contiguous with Xizang plateau on the northern side, the S.E. China and upper Myanmar mountainous parts on the north eastern borders. The forested hilly tracts in the 7 states of N.E. India, largely separated from the eastern Himalaya by the mighty Brahmaputra river and the Assam plains, account for about 10 taxa of which 6 are endemics to smaller areas. Only one subspecies, *R. arboreum* subsp. *nilagiricum* is found in the Nilgiris and in near by high ranges of S.W. Ghats in Peninsular India. Further down *R. arboreum* subsp. *zeylanicum* is again a lone migrant to the Central Highlands of Sri Lanka. There are no indications or fossil evidences on the distribution of the genus in the high ranges of the Vindhyas and Satpuras that lie in between the Himalayas and the South Indian Hill tops or in the chains of high hilly ranges viz., the Eastern and the Western Ghats that border the east and west coasts of Peninsular India.

An analysis of the region-wise/area-wise/state-wise distribution of taxa in India is tabulated below.

Table 1 Distribution of the genus Rhododendron in India

Biogeo-graphic Zone	**Western Himalaya**			**Eastern Himalaya**			**North East India**				**Peninsular India**
State	**J&K.**	**Himachal.**	**Uttaranchal.**	**Sikkim**	**Darjeeling**	**A.P**	**Meghal.**	**Manipur.**	**Nagal.**	**Mizo.**	**Nilgiris**
Species	6	5	5	34	13	75	1	5	2	2	0
Subsp.	1	1	1	1	0	4	0	1	0	0	1
Vars.	0	0	0	2	2	7	2	3	1	0	0

It should be borne in mind that many species are common to some or most of the distribution areas / states in the above 3 geographical regions or biogeographic zones.

5. Diversity and Ecology

The species of *Rhdodendron* exhibit considerable diversity and variation in their habitat requirements, altitude, habit, hardiness, scales and indumentum, inflorescence, flower size, shape and colour and in some other aspects. They range from small prostrate shrublets reaching from 1 to 1.5 m in length with tiny leaves measuring 3 - 6 × 2 - 4 mm as in *R. nivale* subsp. *nivale*; to shrubs of 2 - 6 m high, and trees up to 30 m tall with leaves up to 1 m length as in the case of *R. sinogrande; R. grande* and *R. falconeri* to a lesser degree. Many of the species are shrubs or small trees. The range of colour variation in the same species is also wide in some cases.

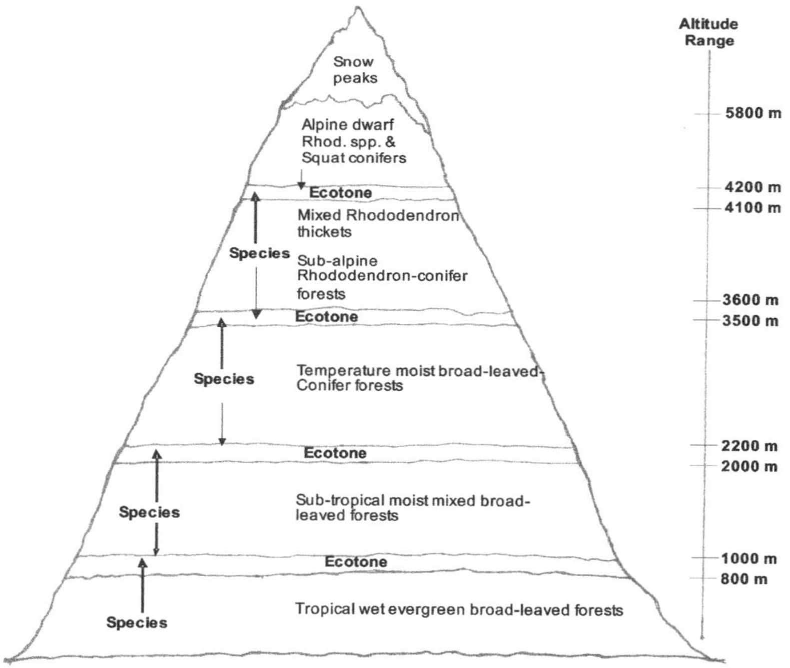

Some representative Rhododendrons

R. nivale ssp. nivale

setosum, lepidotum, chaemaethomsonii, fulgens, campanulatum, campylocarpum, calostrotum

thomsonii, campanulatum, camelliflorum, wallichii, crinigerum, falconeri, cinnabarinum, collettianum, cephalanthum, lanatum, baileyi, barbatum, calostrotum, maddenii, vaccinioides, anthopogon, keysii, fulvum, hookeri, hodgsonii, lepidotum, ciliatum, barbatum

anthopogon, hodgsonii, arizelum, vaccinioides, wightii, beenianum, calostrotum, camelliiflorum, lindleyi, campanulatum, arboreum, campylocarpum, concinnoides, dalhousiae, subansiriense, edgeworthii, santapaui, macabeanum, wattii, arb.ssp delavayi var. peramoenum, kendrickii, ciliatum, barbatum, elliottii, falconeri, arboreum, arb. ssp. cinnamomeum, dalhousiae, grande, griffithianum, hookeri, boothii, coxianum, johnstoneanum, walongense, xanthostephanum

arboreum, formosum var. inaequale, dendricola, nuttallii, veitchianum etc.

arboreum ssp. nilagiricum

arboreum ssp. arboreum

None in India

Diagrammatic representation of some Rhododendron species in different vegetation zones in the Himalayas & N.E. India ; some with wide altitudinal amplitude. Maximum species are found in Temperature-sub-alpine altitudes

Majority of *Rhododendron* species in India are terrestrial; about 15 are either epiphytic on evergreen broad-leaved trees in sub-tropical, sub-temperate and temperate forests growing together with lichens, mosses, ferns, aroids, orchids, *Vaccinium* and *Agapetes* species ; or lithophytic on moss-lichen-covered boulders in alpine screes, meadows and mountain cliffs. They are seen in the altitudes ranging from 1000 - 6000 m with *R. arboreum* first appearing at lower altitudes of 1000 m as noticed at Barapani near Shillong in Meghalaya and elsewhere. It is *R. nivale* subsp. *nivale* that is recorded from the highest ever altitude growing in the snow-line areas between 5800 - 6000 m in Sikkim. Many of the species grow as undergrowth in the altitudes of 2500 - 4500 m in dense forests associated with Oaks, Michelias, Magnolias, Acers, Lawrels, Pines, etc. **(Photo Nos. 1-4).**

Most of them prefer well drained acidic loamy soils laden with humus, while some of the species in the alpine areas grow in sheltered wet gravelly sands along the mountain streams and rivers. Seedlings of *R. arboreum* are usually seen growing along freshly cut, exposed hill palisades and slopes as noticed at Shillong in Meghalaya. To withstand snowy winters they produce tightly overlapping thick scales forming 'leaf-buds' at the apices of branchlets to effectively enclose and protect the inner growth-buds which remain dormant during the harsh snowy winter nights and bright sunny days. The leaves of many species, particularly of those in the open high altitudinal areas, are rolled inwards along their margins to minimize exposure to snow and damage during winter. Besides, many species are clothed with dense indumentum or scales which protect the younger plant parts and reproductive bodies.

Some shrubby species are often found forming large populations at high altitudes commonly called the 'rhododendron thickets' comprising species of *R. hodgsonii, R. anthopogon, R. lepidotum,* etc. in association with conifers like *Juniperus wallichiana, Abies densa, bamboos* and the like. **(Photo No. 11).**

Pollination in most species is carried out by insects. Wild bees, honey bees and butterflies are reported to aid in self and cross-pollination of rhododendrons in Sikkim resulting in the existence of actively evolving hybrid populations at Phuni and Shingba in North-East Sikkim. Bird pollination by honey birds is reported by H. Sleumer (1966) in Malesia and in *R. wightii* by Pradhan and Lachungpa (1990) in Sikkim.

Rhododendron is a polymorphic genus with polyploidy in many species which is a sign of active speciation and continued evolution (Janaki Ammal, 1950).

In most of the species, numerous minute sawdust-like seeds are produced in the capsules that dehisce length-wise to release them and to be effectively dispersed by wind and water to distant places. Seed germination is fairly good giving rise to a number of seedlings. In some species of the subsection *Pseudovireya* the seeds are drawn into long thread-like appendages on either side, eg. *R. santapaui* and *R. vaccinioides*, which aid in dispersal and anchorage for better chances of germination on other trees and survival. It is possible that the seeds being minute with scanty endosperm favour association with mycorrhiza to germinate as in the case of orchids.

Much of the information pertaining to ecology and diversity is known from field observations and recordings of many rhododendron collectors, particularly of F. Kingdon-Ward whose field notes, books and articles are replete with interesting field observations.

6. Economic uses

Rhododendrons are used by local people in the Himalayan region in several ways. Else where they are some of the choicest plants in horticulture. *R. arboreum* is the National Flower of Nepal and *R .niveum* is the State Flower of Sikkim. The Monpas in West Kameng and Tawang districts in Western Arunachal Pradesh use the aromatic twigs and leaves of *R. anthopogon* like Juniper leaves for smoking and as incense. The most important use is for fuel needs. At high altitudes local inhabitants and migrant work-force in Sikkim and Arunachal Pradesh excessively depend on rhododendrons for fire wood needs and heating. Flowers of *R. arboreum* and other brightly coloured species are used for decoration in festivals and religious ceremonies and in worship in Buddhist Monastries. Fresh flowers of *R. arboreum* are used in making jams and are specially brewed to make a pleasant wine in Sikkim and Darjeeling region, and its hard, close grained wood is used for making gift items and small boxes, etc. The leaves of *R. arboreum* and *R. cinnabarinum* are reported to be poisonous to sheep. Pradhan and Lachungpa (1990) enumerate some other interesting details on uses of rhododendrons.

7. Threat factors and conservation status

Extensive natural and anthropogenic alterations of rhododendron habitats and excessive dependence on some of the species for fuelwood purposes **(Photo No. 13)** are the main factors that have rendered many of them to become rare and vulnerable and loose their presence in several places in their distribution range (Sastry & Hajra, 1983 ; Mao et. al., 1990). Some species are naturally rare and sporadic and endemic in their occurrence with niche preference as in the case of epiphytic species. Road building activities, slash & burn agricultural practices, clear felling of large associated trees in rhododendron-mixed forets have brought about ecological degradation of habitats and consequent vulnerability to many rhododendrons. Natural calamities like deficient or excessive rain fall, hot and prolonged summers and snowy winters, land slides also take a toll of species that grow especially in open high altitude areas. These adverse factors negatively influence the germination rate of seeds and regeneration of species.

Rhododendrons also suffer from some pathogenic fungal diseases. 10 rust fungi have been reported on the leaves of many Asiatic species (Leppik, 1974), which in times of virulence can cause severe defoliation and damage to the plants.

The age-old Sacred Groves and some recently established Biosphere Reserves, National Parks and Wildlife Sanctuaries in the *Rhododendron* distribution areas have no doubt laid some restrictions on the destruction of forest areas. This helped the constituent biological species, particularly some animal species in comparision to plant species which are steadily depleting even without being noticed.

Though many species of *Rhododendron* are now in cultivation particularly in the West, little care is taken with regard to the speices in the wild and their habitats in India. The Sikkim Forest Department's initiative to demarcate and fence certain rhododendron rich sites between Lachung and Yumuthang in North-East Sikkim **(Photo No. 2)** and at Varsay in the West district and designate them as 'Rhododendron Sanctuaries' is a laudable step for *in situ* conservation of some species. The Shingba sanctuary also protects some species. **(Photo No. 12).**

Many of the *Rhododendron* species can be grown in the Himalayan region in the altitudes of 1500 to 2500 m except for some species like *R. arboreum, R.dalhousie, R.maddenii* and a few such low altitude preferring species. They survive well in sheltered areas with well drained, humus rich, moisture holding acidic soils having pH in the range of 4.5 to 5.5 under the shade of other broad-leaved forest trees which offer desired filtered sunlight **(Photo No. 7).** They can be propagated through seeds and cuttings which when germinate and take roots can be transplanted during the onset of rains after about an year's growth under care **(Photo No. 8).** Pradhan and Lachungpa (1990) give details on the subject.

There is also the need to raise plantations of other fast growing species in the high altitude areas as alternative fuelwood resource to minimise cutting of naturally growing rhododendrons. It is also necessary to undertake ecological studies on rhododendrons for better understanding of their conservation needs. Establishment of arboreta of *Rhododendron* species at suitable altitudes and species rich areas in Sikkim and Arunachal Pradesh would help further studies and educating the people to conserve the species.

8. The journey covered

The botanical trekkings undertaken during the 1960s by the P.I. (A.R.K.S) in the unexplored parts of Subansiri Division in the erstwhile North Eastern Frontier Agency which encompasses the present political state of Arunachal Pradesh interested him in the study of the genus in India. It was however, during his second botanical exploration in the subtropical, temperate and subalpine forested areas lying along the S.E mountain ridges of Subansiri division in the altitudes of 1,660 – 3,500 m (ca.5,000 – 10,500 ft) in April-May 1965 jointly with Mr.Peter Cox, Sir Peter Hutchinson and Mrs. Patricia Cox, **(Photo No. 5)** provided an opportunity of closely observing and collecting as many as 20 different spectacular species of the genus that were in full bloom. During this exploration Peter Cox & Peter Hutchinson helped him to get basic knowledge on the field identification and presented a copy of Rhododendron Year Book (1963). In this tour *Rhododendron kendrickii* and *R. nuttallii* earlier collected from Bhutan and in the Western Kameng district of Arunachal Pradesh; *R. arizelum, R. peramoenum* (now treated as *R. arboreum* subsp. *delavayi* var. *peramoenum*), *R. stenaulum* (now treated under *R. moulmainense*) were collected, studied and reported as extended distributional records for India with their earlier distributions known only in far off provinces in China. Besides these, nearly 4 to 5 species of the genus that were collected during this joint botanical exploration and in others turned out to be new to science. Of these, epiphytic saplings of a *Rhododendron* species collected by the Principal Investigator (and also Peter Cox) were successfully transplanted and grown in the Botanical gardens of the Eastern Circle of B.S.I. at Shillong and Glendoick gardens of the Coxes near Perth in Scotland, U.K. The Shillong plants first flowered in April 1966 making it possible to critically study, and publish it as a new species under the name *R. santapaui*.

Photo No. 1. Temperate forests with *R. grande* in Subansiri.

Photo No. 2. Rhododendron sanctuary at Yumthang, North Sikkim.

Photo No. 3. Temperate forests with rhododendron undergrowth.

Photo No. 4. *R. thomsonii* bushes in Yumthang valley, North Sikkim.

Photo No. 5. Peter & Patricia Cox, Hutchison with A.R.K.S in Subansiri in April, 1965.

Photo No. 6. Rai & A.R.K.S in Sikkim.

Photo No. 7. Seeddling of *R. hodgsonii*.

Photo No. 8. Root system of *R. hodgsonii* seedling.

Photo No. 9. Unfolding of leafbuds in *R. falconeri*.

Photo No. 10. *R. hodgsonii* with mature fruits.

Photo No. 11. Rhododendron thicket.

Photo No. 12. Shingba Rhododendron Sanctuary.

Photo No. 13. Rhododendron wood stacked at Se La Pass, Kameng, Arunachal Pradesh.

Photo No. 14. A.R.K.S besides *R. campylocarpum* bush, holding flowering twigs of *R. campylocarpum* & *R. cinnabarinum* in North Sikkim.

Photo No. 15. *R.santapaui* growing in Edinburgh Botanical Garden.

Photo No. 16. Type specimen of *R. santapaui* in CAL.

In the later years, the present authors while working in the U.S. Fish & Wildlife funded Botanical Survey of India (BSI) project on the Study, Survey and Conservation of Endangered Species of Indian Flora (POSSCEF) have published a paper on the Rare, Endangered and Endemic Species in India and made a plea for their conservation by identifying certain Rhododendron rich sites in Sikkim and Arunachal Pradesh (Sastry & Hajra, 1983). They have also written datasheets for some rare, threatened and endemic species included in the first volume of Plant Red Data Book–I (Jain & Sastry, 1984) and in the Plant Conservation Bulletins broughtout under the POSSCEF project between 1980 to 1985. During April-May, 1978, both the authors of this book have undertaken botanical survey work in North Sikkim for collection of *Rhododendron* species and other threatened plants in the flora of Sikkim. Also, some parts of Meghalaya, Mizoram have been surveyed and notes on the Conservation status of some species were recorded in the field and from the historical herbarium collections deposited in the Kanjilal and Central National Herbaria of B.S.I. at Shillong and Calcutta respectively, and Llyods Botanical Garden at Darjeeling.

In addition to these efforts, in May, 1993 the P.I. led a multi-disciplinary team with scientists from the Botanical, Zoological, Anthropological Surveys located at Calcutta which extensively toured in Sikkim following the historical trails of Sir J.D. Hooker made during the years 1848-1850. Particular attention was paid to assess present conservation status of *Rhododendron* species reported by Hooker which revealed changes in their population status and habitats. **(Photo No. 6).**

Historical and recent collections present in the B.S.I. Herbaria at DehraDun, Shillong and Calcutta have been examined and listed, with notes on their distribution, etc. The P.I. during his tenure as Project Coordinator of the endangered species programme visited the U.S. Herbaria in the Smithsonian Institution, Arnold Arboretum, New York Botanical Garden, Berkely University, St. Louis Botanical Garden and the Kew Herbarium of the Royal Botanic Garden, U.K. during July–August,1981 and made notes on the Himalayan *Rhododendron* collections for the project work.

All this data was being compiled to make a meaningful account on the taxonomy and distribution of the genus in India but could not make a headway towards completion due to other departmental assignments and responsibilities. Finally the time had drawn close to superannuation of the P.I. from B.S.I in January 1996. His post-retirement assignments as Director of Biodiversity Hotspots Conservation Programme and Biodiversity Conservation Prioritization Project in WWF-India Secretariat at New Delhi up to March 2001 kept him fully engaged. In 2002, during an informal meeting and discussions of the P.I. with Dr. Prof. Rolla S. Rao (who retired as Director-in-charge, B.S.I. under whom the P.I. joined the Survey at Pune in the year 1961) the unfinished subject came up and at his behest and encouragement, the senior author (A.R.K.S) submitted a project proposal to the Department of Science & Technology, Govt. of India under its USERS (Utilization of Scientific Expertise of Retired Scientists) scheme through the Society for Conservation and Management of Natural Resources, a registered NGO at Hyderabad with Prof. Rolla S. Rao as its Hony. President. The project was granted in March 2004. As the preparations were on to resume the project work, the Principal Investigator and his wife had to go to the U.K. for certain family obligations. During the six month long stay with his son-in-law and daughter in London from April-September 2004, he visited the Kew Herbarium and the Library of the Royal Botanic Garden and collected authentic herbarium data and important literature on the genus in India. He also visited the Herbarium and Library of the Royal Botanic Garden at Edinburgh, Scotland and collected latest taxonomic references, examined and photographed the Type specimens of several Indian rhododendrons which are housed in this herbarium. Further, he held discussions with experts on the genus *Rhododendron*, namely, Drs. D.F Chamberlain and George Argent who provided very useful suggestions and some literature. The Principal Investigator also made a visit to Glendoick to meet and renew old friendship and association with Mr. Peter Cox, Mrs. Patricia Cox and their son Kenneth Cox who are the authors of the monumental book : *The Encyclopedia of Rhododendron Species.* The Coxes not only gave valuable suggestions for writing the present book but also presented a copy of their most valuable publication noted above, besides a very warm, touhing and memorable hospitality at their sprawling Villa in Glencarse and have shown round their impressive living collections of several hundred species of *Rhododendron* introduced from the Sino-Himalayan region and other parts of the globe.

The Herbarium data collected, photographs of the Type specimens of Indian taxa by the P.I. and some latest taxonomic literature presented by the Scientists of the Edinburgh and Kew Herbaria proved to be of immense value in finalizing this book. Also, visits to the B.S.I Regional Herbaria at DehraDun, Coimbatore and the Central National

Herbarium (CAL) at Sibpur near Calcutta were undertaken by the P.I. during the project work.

It is hoped that the book will be of help in the identification of our natural heritage of the magnificient Himalayan rhododendrons in India and will be of educational value to students and teachers, nature lovers, foresters and plant conservationists in India. It should also be of help to undertake further research in ecology, propagation and conservation of several *Rhododendron* species that are becoming rare, vulnerable and endangered due to various factors and in restoration of their degraded habitats and populations in the fragile mountain ecosystems of the Indian Himalayan region and other mountain ranges in North East India.

Leaf shapes of Rhododendrons

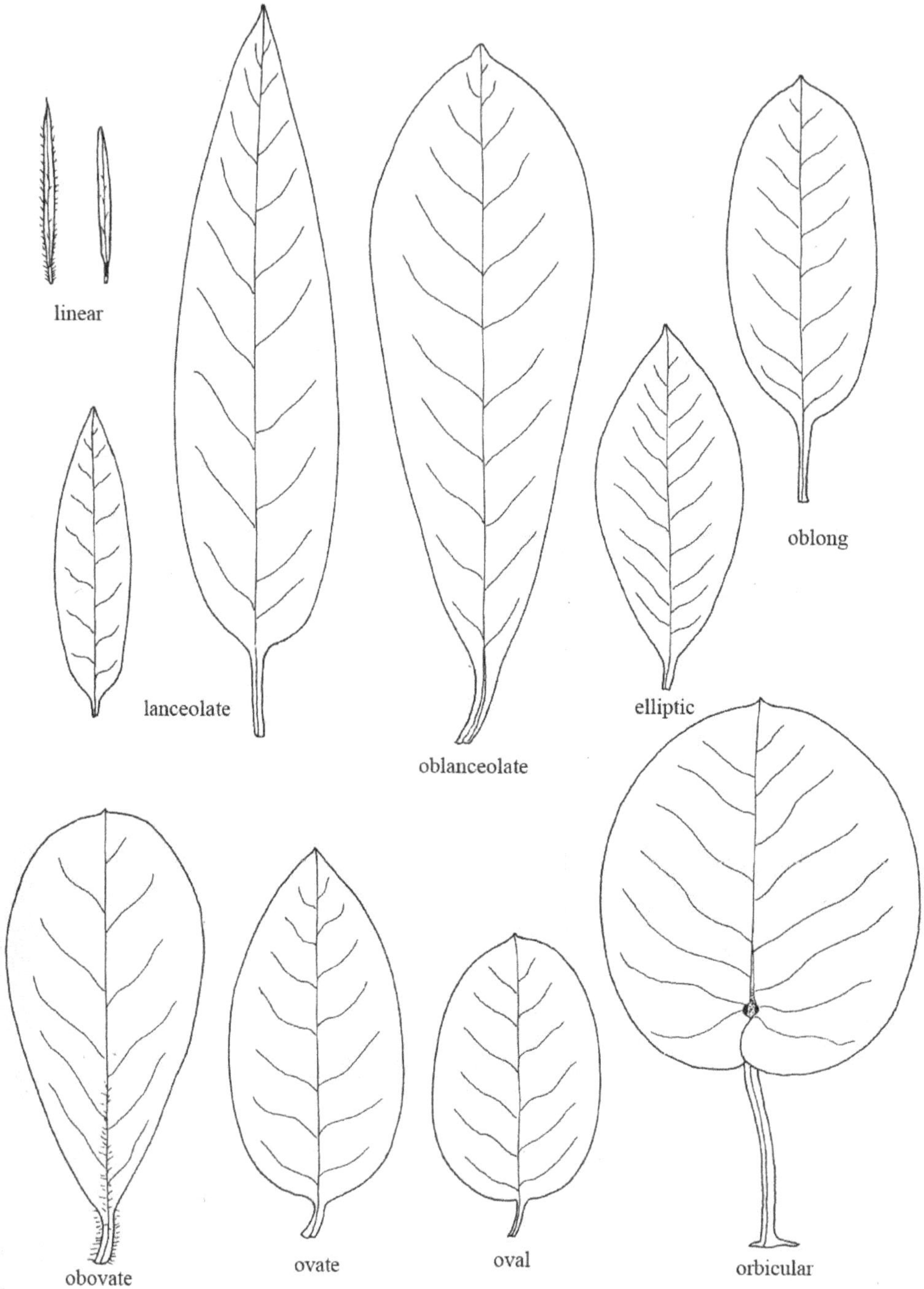

Source: Davidian, H.H., The Rhododendron Species, Vol.I, 1982

Rhododendron Flower Shapes

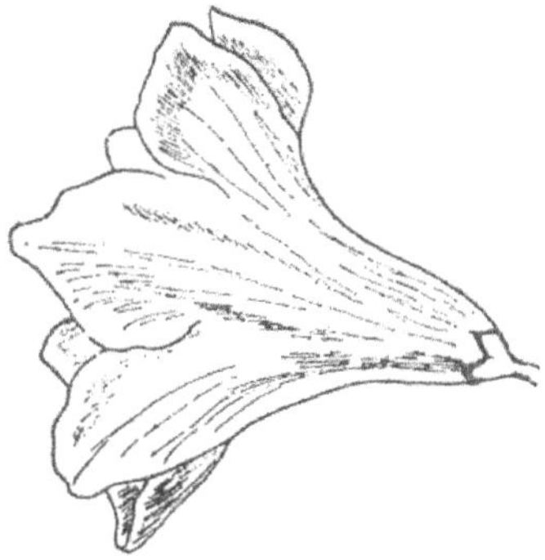

Funnel-shaped

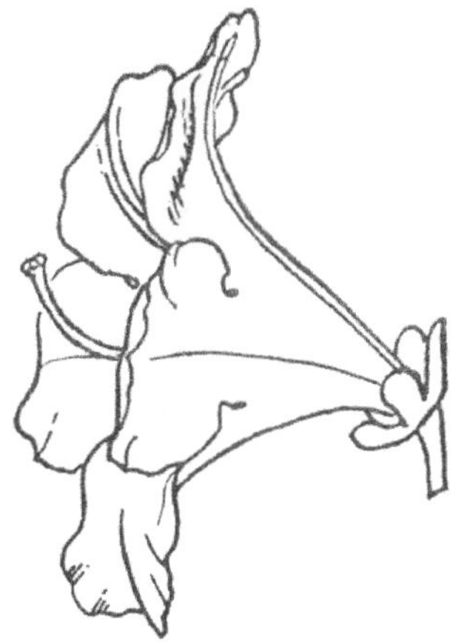

Openly funnel-shaped

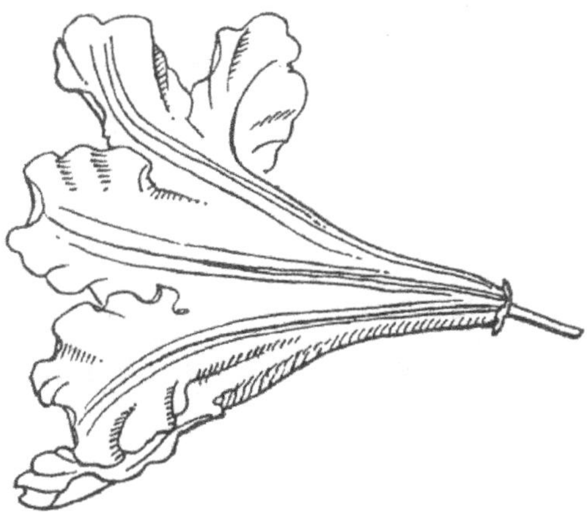

Open-funnel-campanulate

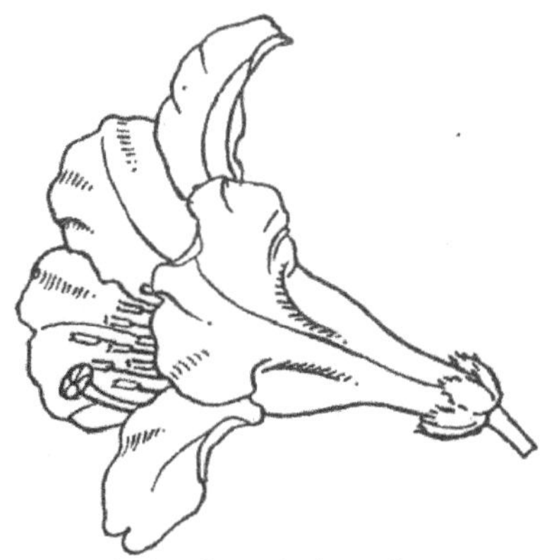

Tubular funnel-shaped

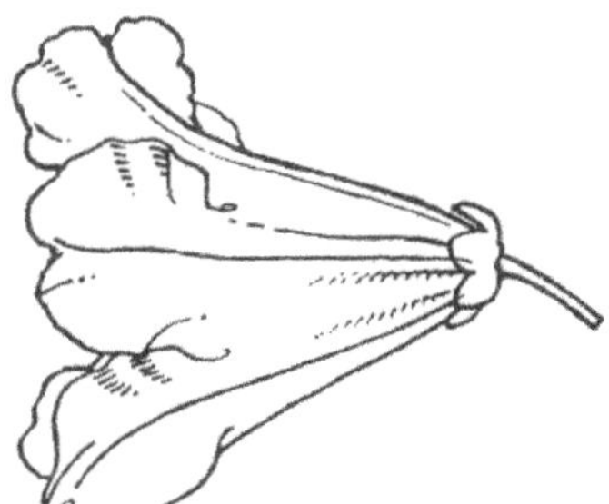

Widely funnel-campanulate

Widely funnel-shaped

Source: Rhododendron Year Book, 1963. RHS, London

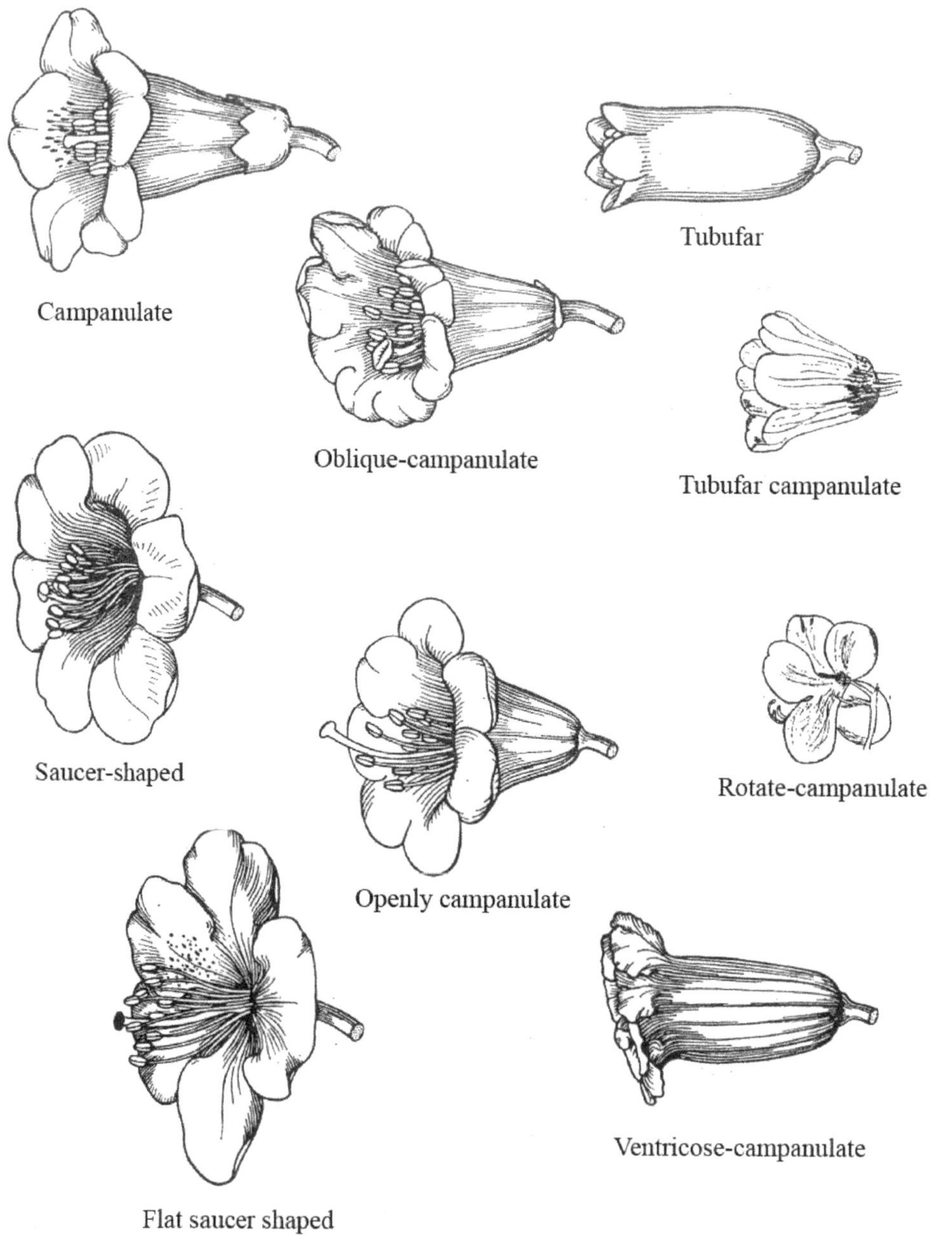

Source: Rhododendron Year Book, 1963. RHS, London

Scales

Entire: *R. crassum*

× 125
viewed from above

Undulate: *R. impeditum*

× 137
viewed from above

Lacerate: *R. anthopogon*

× 125
viewed from above

Crenulate: *R. saluenense*

× 125
viewed from above

Vesicular: *R. trichocladum*

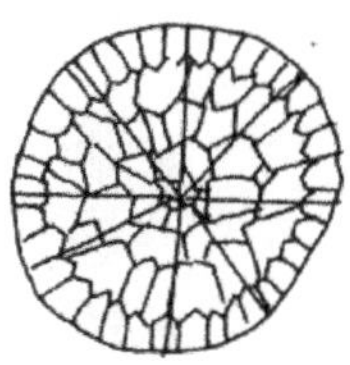

× 125
viewed from above

R. trichocladum

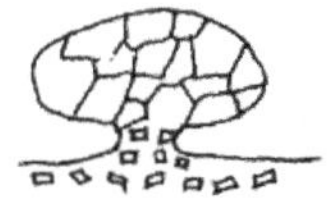

× 80
viewed from side

Source: Davidian, H.H., The Rhododendron Species, Vol.I, 1982

Seeds

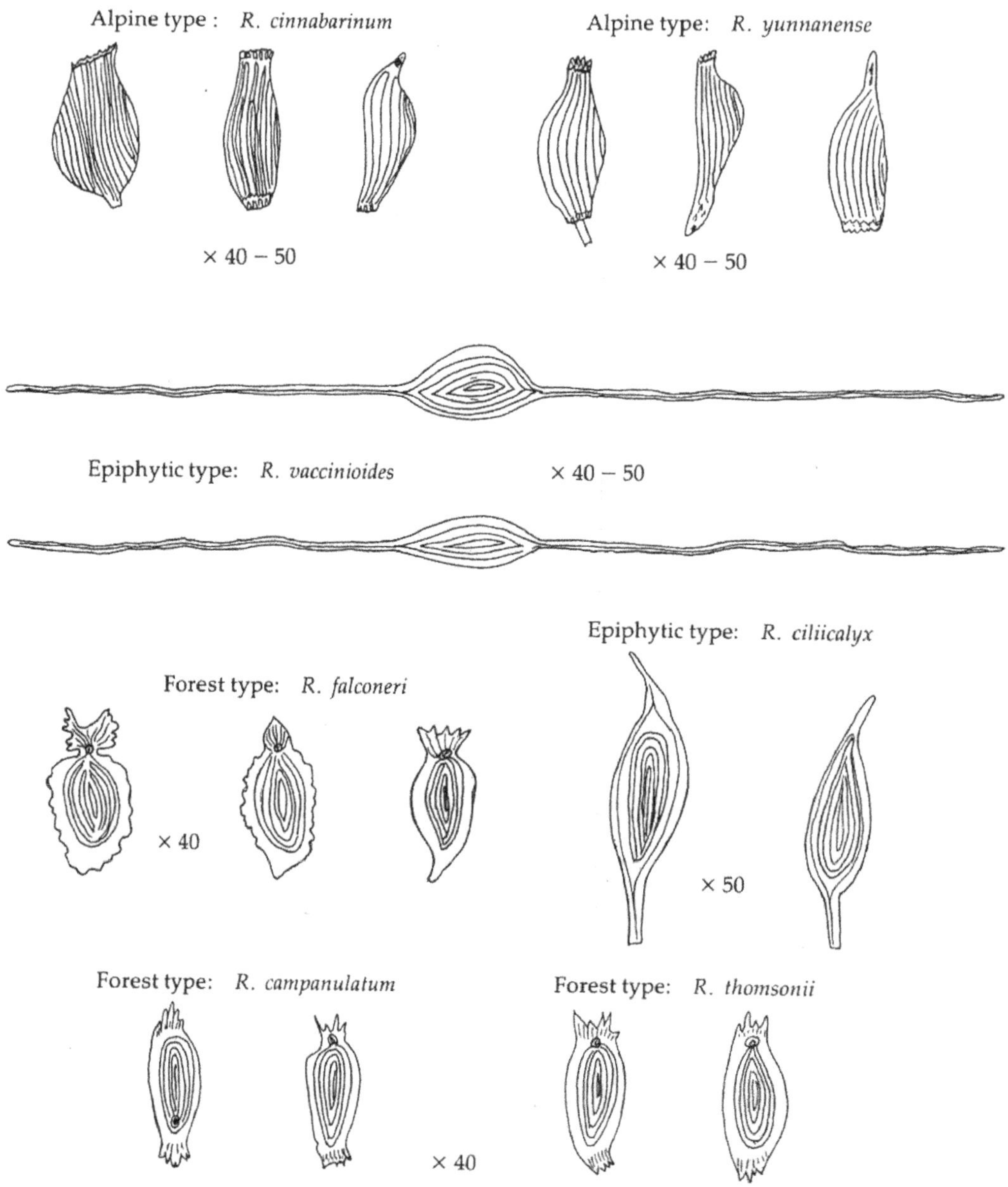

Source: Davidian, H.H., The Rhododendron Species, Vol.I, 1982

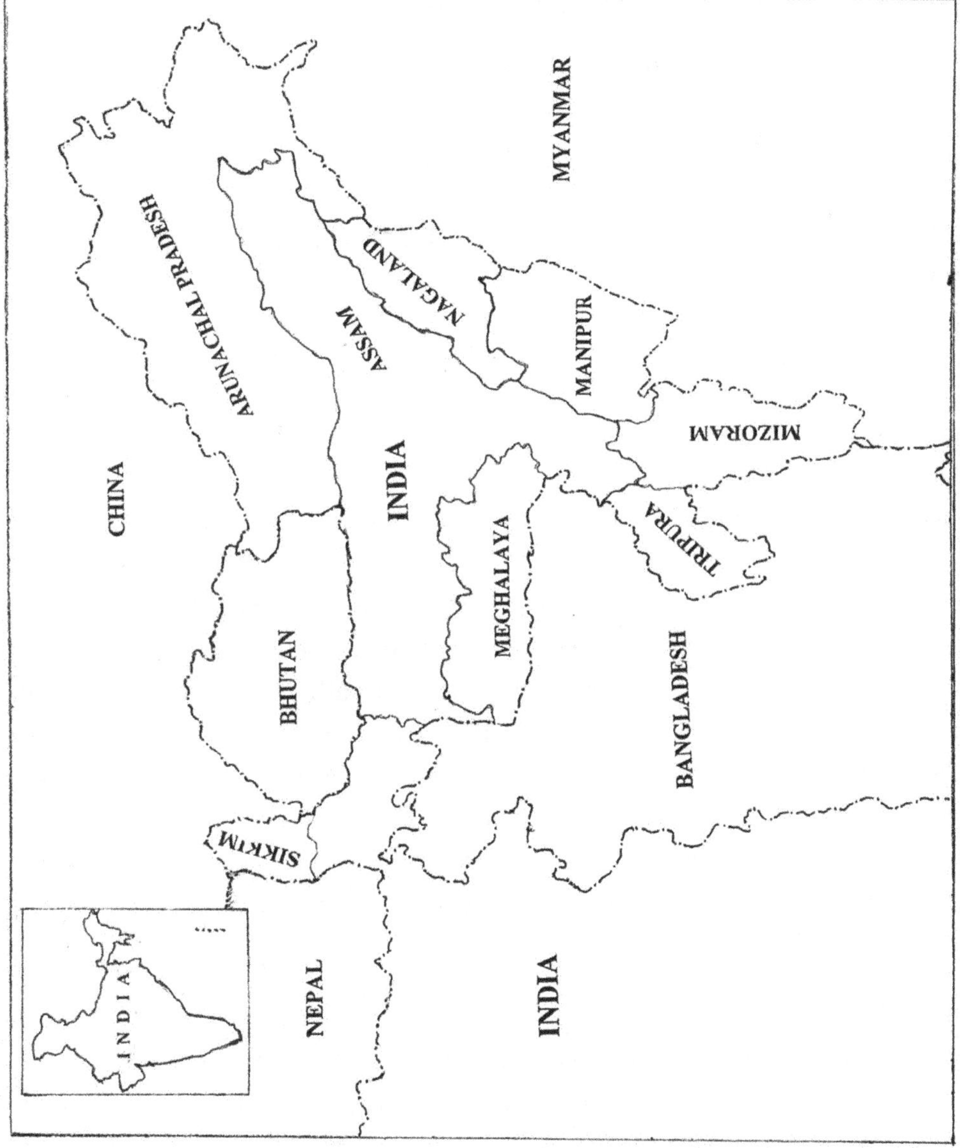

Map 1. **N.E. Indian States & adjoining countries**

(Maps are only indicative and not to scale)

2
Taxonomic Treatment of the Genus

Rhododendron L.

Rhododendron L., Species Plantarum ed.1,392(1753). C.B. Clarke in Hook.f., Fl.Brit. India 3:462 (1882).

Azalea L., Species Plantarum ed.1.(1753).

Terrestrial or sometimes epiphytic shrubs, trees or sometimes creeping shrublets. Plant parts glabrous or often covered with scales or indumentum consisting of various types of hairs. Leaves alternate, often clustered at the ends of branches, rarely subopposite or in psuedowhorls, usually entire, leathery. Flowers fascicled or subcorymobose, terminal or axillary, occasionally solitary. Bracts broad, generally caducous. Bracteoles linear. Calyx 5-lobed, conspicuous or small or obsolete, rarely saucer-shaped, persistent. Corolla campanulate, widely funnel-shaped or rotate or cylindric, tube long or short, lobes 5 - 10. Stamens (4) 5 – 20 (27 in *Maddenia* subsection), usually 10; anthers large, oblong, dehiscing by terminal pores. Ovary 5 - 20-celled; style long or short, straight or sharply deflexed; stigma capitate; ovules numerous in each cell. Capsule short or long, woody or membranous, septicidally dehiscing from apex, 4 – 20-valved; valves breaking away from the placentas. Seeds numerous, fusiform or ellipsoid or irregular, albuminous; testa close or loose, often shortly crested or finned or drawn into long tails at the ends.

Type species: *R. ferrugenium* L. (Lectotype).

General distribution: With over a thousand species, the genus is widely distributed in the tropical, sub-tropical forests of Malaysia, Indonesia, Papua New Guinea and in sub-tropical, temperate, sub-alpine and alpine areas in the forested mountains of Asia, Europe and North America. It is not distributed in Africa and South America.

Plan of Taxonomic treatment: The classification of the genus *Rhododendron* (in India) into subgenera, sections and subsections followed in this book is in accordance with the classification given by Drs. David Chamberlain, Roger Hyam, George Argent, Gillian

Fairweather and Kerry S. Walter in their book: *The Genus Rhododendron: Its classification and synonymy* (Royal Botanical Garden, Edinburgh,1996). Of the 8 subgenera recognized, 4 are distributed in India. However, the subsections (wherever present), the species, subspecies and varieties under the different Sections of the four Subgenera, are arranged alphabetically irrespective of their positions in the taxonomic keys for easy reference.

The delimitation of the species and subspecific taxa under the two subgenera: *Rhododendron* and *Hymenanthes* included in this book, to most extent, is on the lines of taxonomic revisions by Dr. J. Cullen and Dr. D.F. Chamberlain published in the *Notes Royal Botanical Garden*, Edinburgh, vol. 39(1):1980 & 39(2): 1982 respectively, and Cox & Cox in their book : *The Encyclopedia of Rhododendron Species* (Glendoick Publishing, 1997).

Key to the subgenera of genus Rhododendron

1.	Plants and other plant parts covered with scales	*Rhododendron*
1a.	Plants and other plant parts not scaly..........................	2
2.	Creeping to upright shrubs; leaves covered with simple or ribbon-like flattened bristles or stiff glandular hairs...........	*Tsutsusi*
2a.	Shrubs or trees; leaves minutely puberulous or glabrous except on the midrib or covered with dense spongy or woolly rusty tomentum	3
3.	Shrubs or small trees; branchlets usually minutely puberulous; inflorescence axillary, single flowered ; corolla broadly campanulate with spreading lobes; calyx lobes small, to 8 mm long; stamens 5 or 10................................	*Azaleastrum*
3a.	Usually trees or large shrubs; branchlets tomentose or glandular; inflorescence usually terminal, many-flowered; corolla campanulate; calyx lobes minute; stamens10	*Hymenanthes*

3
The Genus Rhododendron L. in India and its Classification

Subgenus Rhododendron

Section Vireya

Subsection Pseudovireya

Section Pogonanthum

Section Rhododendron

Subsection Baileya

Subsection Boothia

Subsection Camelliiflora

Subsection Campylogyna

Subsection Cinnabarina

Subsection Edgeworthia

Subsection Glauca

Subsection Lapponica

Subsection Lepidota

Subsection Maddenia

Subsection Monantha

Subsection Saluenensia

Subsection Tephropepla

Subsection Trichoclada

Subsection Triflora

Subsection Uniflora

Subsection Virgata

Subgenus Hymenanthes

Section Ponticum

Subsection Arborea

Subsection Barbata

Subsection Campanulata

Subsection Campylocarpa

Subsection Falconera

Subsection Fortunea

Subsection Fulgensia

Subsection Fulva

Subsection Glischra

Subsection Grandia

Subsection Irrorata

Subsection Lanata

Subsection Neriiflora

Subsection Parishia

Subsection Taliensia

Subsection Thomsonia

Subgenus Tsutsusi

Section Tsutsusi

Subgenus Azaleastrum

Section Choniastrum

4
Subgenus Rhododendron

Subgenus **Rhododendron** L. Species Plantarum ed.1,392(1753).

Syn. : Subgenus *Lepidorrhodium* Koehne, Deutsche Dendrol. 449 (1853).

Subgenus *Eurhododendron* K.Koch, Dendrol. 2:157 (1852).

Section *Lepidorrhodium* (Koehne) Rheder in Bailey, Standard Cycl. Hort. 5:2937 (1916).

Plants bearing scales at least on the young growth and lower surface of the leaves, often also on the leaf upper surface, pedicel, calyx, corolla, ovary and style. Hairs present or absent, when present either filiform-acicular or loriform type; branched or dendroid type hairs are always present on the leaf-bud scales of section *Pogonanthum*.

Type species: *R. ferrugineum* L.

Subgenus *Rhododendron* has been divided into three sections as under:

Key to the sections of subgenus Rhododendron

1. Capsule valves thin and papery, usually twisted on dehiscence; placentas separating as thread-like structures from the central axis as the capsule opens; seeds long caudate-appendaged, the appendages usually longer than the body of the seed........................ *Vireya*

1a. Capsule valves hard and woody at dehiscence; placentas not separating from the central axis; seeds variously winged and finned, rarely caudate-appendaged, when with appendages these are shorter than the body of the seed........................ 2

2. Scales lacerate; corolla hypocrateriform or funnel-hypocrateriform; plants with characteristic pine-apple like smell; hairs fringing the inflorescence bud-scales dendroid .... *Pogonanthum*

2a. Scales entire, crenulate or undulate; corolla usually tubular, rotate or campanulate, very rarely hypocrateriform; plants variously aromatic but not with the smell of pine-apple; hairs fringing the inflorescence bud-scales simple .... *Rhododendron*

Section Vireya

Section **Vireya** (Blume)Copeland, Philip.Journ.Sci.40:136 (1929).Sleumer, Flora Malesiana ser.1,6:474-668(1968).

Subsection **Pseudovireya** (Clarke) Sleumer, Bot. Jahrb. 74:537(1949).

Syn.: *Rhododendron* subgenus *Pseudovireya* C.B.Clarke in Hook.f., Fl. Brit.India 3:464(1882). Series *Vaccinioides.*

Shrubs, to 1.5 m high, usually epiphytic, compact to spreading. Branchlets usually scabrid. Leaves persistent, evergreen, thick and leathery, 5 cm long, 2.4 cm broad, usually in whorls at ends of branchlets, lower surface lepidote. Inflorescence 1 - 7-flowered. Calyx 0.5 - 4.0 cm long. Corolla campanulate, lepidote outside, small, yellow, light pink to white, waxy. Stamens 10. Ovary lepidote; style short. Capsule dehiscent from top downwards; seeds characteristically ending into long tails on either side.

Type species: *R. vaccinioides* Hook.f.

Notes: The section *Vireya* includes about 300 species mostly found in the tropical forests of S.E. Asia, especially New Guinea and Borneo. In India only two species under its subsection *Psuedovireya* are found in the Eastern Himalayan region.

Key to the species of subsection Pseudovireya

1. Leaves obovate-lanceolate, 2.5 - 4 cm long, 1.0 - 1.5 cm broad; corolla 1.2 - 1.3 cm long, creamy white, waxy with red scales outside .... *santapaui*

1a. Leaves spathulate-obovate, 0.5 - 2.5 cm long, 0.4 - 1 cm broad; corolla 0.6 - 1.1 cm long, white to lilac pink, not waxy .... *vaccinioides*

R. santapaui A.R.K. Sastry, S.K. Kataki, P.A. & E.P. Cox, P.C. Hutchison in Journ. Bombay Nat. Hist. Soc. 63(3) : 742 – 747 (1968). **(Photo No.s 15 & 16).**

Epiphytic twiggy shrub, to 1.5 m high. Branchlets scabrid; young shoots densely scaly. Leaves subverticillate, 2.5 - 4 × 1.1 - 1.5 cm, thick and fleshy, obovate-lanceolate, upper surface dark green, lower surface pale, scaly, scales unequal, brown, 2 - 3 x their own diameter apart. Inflorescence terminal, 2 - 4-flowered. Calyx 1 mm, densely scaly.

Corolla campanulate, 1.5 - 2.5 cm long, white to creamy-white, waxy with red punctate scales outside. Stamens 10, shorter than the corolla. Ovary densely scaly; style bent, shorter than corolla, stamens and ovary. Capsule slender, dehiscing from top, valves twisting upwards; seeds long-tailed at both ends.

Distribution and habitat: India: Arunachal Pradesh (Subansiri District). Epiphytic on large trees in subtemperate forests at 1600 m alt. Possibly endemic. **Map 2.**

Notes: This is a rare and distinctive species collected by the authors in a joint Boatnaical exploration in April, 1965. The vegetative specimens with previous year's capsules reared in the 'Woodlands compound' of the Eastern Circle, Botanical Survey of India, Shillong (by A.R.K.S) flowered during September, 1965 based on which the species was described as new and named in honour of Rev. Fr. H. Santapau, a former Director of the BSI.

R. vaccinioides Hook.f., Rhododendrons of Sikkim Himalaya 3:1849. C.B. Clarke in Hook.f., Fl. Brit. Ind. 3:464 (1882).

Epiphytic straggling shrub, to 1 m high. Branchlets densely scabrid, many-leaved. Leaves 0.5 - 2.5 cm long, 0.4 - 1 cm broad, spathulate-obovate to obovate, upper surface smooth, shiny, lower surface with unequal scales, widely spaced, 2 - 5 x their own diameter apart, apex notched, tapering at base to a short winged petiole. Inflorescence 1 - 2 (- 3)-flowered, terminal. Calyx 1 - 3 mm long, 5-lobed, lobes sparsely lepidote. Corolla campanulate, 0.6 - 1.2 cm long, 5-lobed, lobes spreading, thick, white to lilac-pink, lepidote outside. Stamens 10, as long as corolla or slightly exerted. Ovary and style glabrous. Capsule 2.0 - 2.5 cm long, slender, lepidote, valves membranous, recurved to their base on dehiscing; seeds with long tails at both ends.

Distribution and habitat: India : Sikkim, Darjeeling, Arunachal Pradesh. Nepal, Bhutan, China (S.E. Xizang, N.W. & W. Yunnan), N. Myanmar. Inhabits moist mossy rocks and shady, moss-covered trees in the altitudes of 1800-4300 m. **Map 2.**

Notes: The species epithet denotes the 'Vaccinium-like Rhododendron', for its leaves resemble those of *Vaccinium obovatum.*

Section Pogonanthum

Section **Pogonanthum** G. Don, Gen. Hist. Dichlam. Pl. 3:843 (1834). Cullen, Notes R.B.G. Edinb. 39(1) : 156(1980). Cox & Cox, The Encyclopedia of Rhododendron Species 225 (1997).

Syn.: Series *Anthopogon & Cephalanthum* sensu Hutchinson in Stevenson (ed.) The Species of Rhododendron 4 & 198 (1930).

Series *Anthopogon* sensu Cowan & Davidian, Rhodo. Yearbook 2:64 (1947).

Small shrubs or undershrubs, twiggy or sometimes prostrate, *ca* 1 m high, with characteristic aroma of pine-apple. Leaves evergreen, strongly aromatic, upper surface dark green, lower surface densely lepidote, scales overlapping and lacerate. Leaf-bud scales deciduous or persistent. Inflorescence terminal, dense, a many-flowered clustered racemose umbel. Flower-bud scales always fringed with large, branched, tree-like (dendroid) hairs. Calyx distinctly 5-lobed. Corolla narrowly-tubular with spreading lobes, white, pink, cream or yellow; corolla tube with a ring of hairs in the throat. Stamens 5 - 10, included in the corolla tube, filaments puberulous at the base. Ovary lepidote, 5(- 4)-celled; style very short, club-shaped, included, stigma truncate. Capsule short, ovoid. Seeds unwinged.

Type species : *R. anthopogon* D.Don

Key to the species of section Pogonanthum

1. Leaf-bud scales very early deciduous; leaves broadly lanceolate, upper surface pale green, lower surface light yellowish-buff; corolla funnel-hypocrateriform; stamens 8 - 10. .... *collettianum*

1a. Leaf-bud scales persistent or deciduous, very conspicuous; leaves oblong-elliptic to ovate, upper surface dark green, lower surface light yellowish to reddish brown; corolla narrowly tubular; stamens usually 6 - 8 2

2. Fairly compact shrubs; leaves glossy green above; calyx with ciliate margins; inflorescence flat topped to rounded .... *cephalanthum*

2a. Prostrate or upright shrubs; leaves dark green above; calyx lepidote and hairy; inflorescence open topped .... *anthopogon*

R. anthopogon D. Don, Mem. Wern. Soc. 3:409 (1821). C.B. Clarke in Hook.f., Fl. Brit. India 3:472 (1882). Cullen, Notes R.B.G. Edinb. 39(1) : 158(1980). Cox & Cox, The Encyclopedia of Rhododendron Species 225 (1997). **(Photo No.s 17 & 18).**

Shrub, to 1.5 m tall. Leaf-bud scales persistent or not. Leaves very aromatic, 1 - 3.8 x 0.8 - 2.5 cm, subacute, mucronate or slightly emarginate at apex, lower surface reddish-brown, densely scaly, scales overlapping, in 2 - 3 tiers. Inflorescence 5 - 9-flowered; pedicels short, lepidote. Calyx lobes oblong, 3.5 - 5.0 mm, lepidote and ciliate on margins. Corolla 0.8 - 1.6 cm long, white, pale yellow, pink to deep rose, tube densely pilose inside. Ovary lepidote. Capsule 4 - 5 mm, lepidote, slightly longer than the calyx.

Two subspecies are recognized the populations of which are predominant in two geographic ranges, though sometimes are found growing together.

1. Leaf-bud scales deciduous .... subsp.*anthopogon*

1a. Leaf-bud scales persistent .... subsp.*hypenanthum*

subsp. **anthopogon**

Syn.: *R. haemonium* Balf.f. & Cooper, Notes R.B.G. Edinb. 9:283 (1916). *R. anthopogon* D. Don var. *haemonium* (Balf.f. & Cooper) Cowan & Davidian, Rhodo. Yearbook 2:68 (1947).

Distribution and habitat: India : Western Himalaya-Uttar Pradesh; Eastern Himalaya-Darjeeling mountains, Sikkim, Tawang & Kameng districts in W. Arunachal Pradesh. Bhutan, China (S.E. Xizang). Frequent, in open hill slopes and on rocks in the altitudes of 2,700 - 4,900 m. **Map 2.**

Notes: The epithet *anthopogon* means with 'bearded' flowers indicating the hairs in the corolla tube.

R. anthopogon D. Don subsp. **hypenanthum** (Balf.f.) Cullen, Notes R.B.G. Edinb. 37:327 (1979).

Syn.: *R. hypenanthum* Balf.f., Notes R.B.G. Edinb. 9:291 (1916). Royle, Ill. Bot. Hiamal. t. 64 (1839).

R. anthopogon D.Don

Distribution and habitat: India: Western Himalaya-Kashmir & Uttar Pradesh. Nepal, Bhutan. Frequent, in open slopes and ledges, sometimes in open forest floors in the altitudes from 3,400 to 5,500 m. **Map 2.**

R. cephalanthum Franchet, Bull. Soc. Fr. 32:9 (1885). Cullen, Notes R.B.G. Edinb. 39(1): 161(1980). Cox & Cox, The Encyclopedia of Rhododendron Species 226 (1997).

Prostrate or erect shrub, 0.1 - 1.2 m high. Leaf-bud scales persistent, very conspicuous. Leaves 1.5 - 4.7 x 0.5 - 2.3 cm, margins recurved, rounded at base, obtuse or rounded at apex, scales on the lower surface dense, overlapping, in 2 - 3 tiers, those of the lower tier golden-yellow, paler than those of the upper tiers. Inflorescence clustered, many-flowered (up to 10); pedicels short, lepidote. Calyx lobes oblong, 4 - 7 mm, ciliate on the margins. Corolla white or pink, rarely yellowish, tube 0.6 - 1.4 cm long, glabrous and elepidote outside, densely pilose within, lobes 4 - 8 mm, sparsely scaly on the backside. Stamens 5(- 8). Ovary lepidote. Capsule as long as or slightly shorter than the calyx.

Two subspecies *viz* subsp. *cephalanthum* and subsp. *platyphyllum* are distinguished based on the size of the leaves and corolla-tube length. The two subspecies have also separate geographical distribution ranges.

Subspecies *cephalanthum* which occurs in India has smaller leaves (1 - 2.6 x 0.5 - 1.5 cm) and shorter corolla tube (0.6 - 1.3 cm long).

subsp. **cephalanthum**

Syn.: *R. crebreflorum* Hutchinson & Kingdon-Ward, Notes R.B.G. Edinb. 16:173 (1931). Type: Kingdon-Ward 337, Delei valley, Arunachal Pradesh.

Distribution and habitat: India : E. Arunachal Pradesh, Delei valley, Mishmee Hills. N. Myanmar, China (N., N.W. & E. Yunnan, S. & S.E. Xizang). Rare; perhaps known only from Kingdon-Ward's collections along rocky slopes and on mountain cliffs at altitudes from 2,700 - 4,900 m. **Map 2.**

Notes: The species name *cephalanthum* means with flowers in a head.

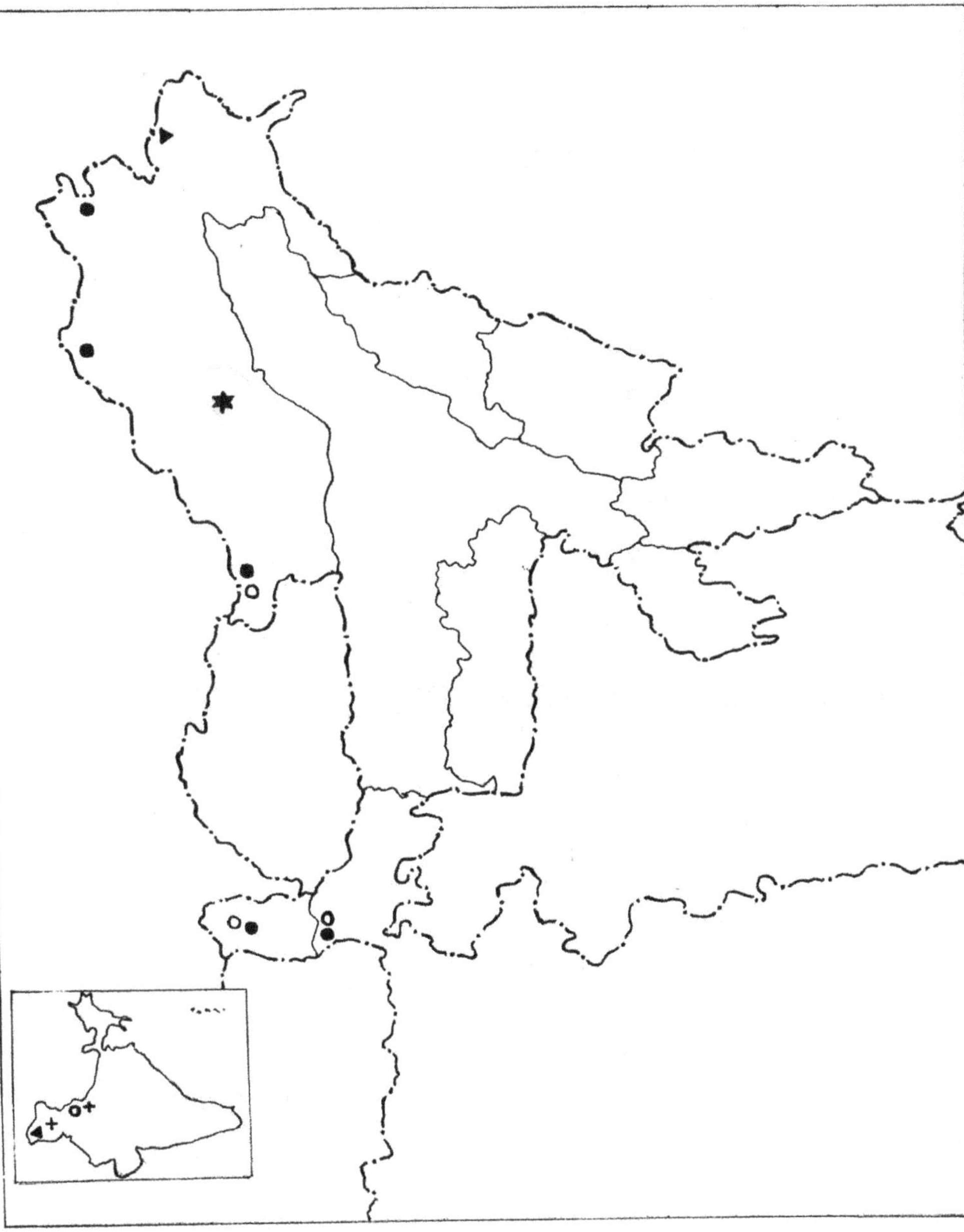

Map 2. ★ R.santapaui. ● R.vaccinioides. ▲ R.collettianum. ▼ R.cephalanthum. O R.anthopogon ssp.anthopogon. + ssp.hypenanthum.

(Maps are only indicative and not to scale)

R. collettianum Aitchison & Hemsley, Journ. Linn. Soc. 18:75 (1881). Cullen, Notes R.B.G. Edinb. 39(1):158(1980). Cox & Cox, The Encyclopedia of Rhododendron Species 228 (1997).

Shrub, to 1 m high. Leaf-bud scales deciduous. Leaves 3 - 6 x 1.3 - 1.7 cm, broadly lanceolate, margins slightly recurved, base rounded, apex acute with a small mucro, upper surface pale green, elepidote, lower surface light yellowish-buff with overlapping, plastered, golden-brown scales. Inflorescence 5 - 12-flowered (rarely more). Calyx lobes 3 - 5.5 mm, sparsely lepidote, loriform-setose along the margins, puberulous within. Corolla narrowly tubular funnel-shaped, 1.5 - 2.4 cm long, white (sometimes pink in bud), glabrous and elepidote outside, pilose within the tube. Stamens 8 - 10. Ovary lepidote. Capsule lepidote, *ca* 5 mm, almost equalling the calyx.

Distribution and habitat: India : N.W. Kashmir. Pakistan, Afghanistan. In steep rocky and stony slopes and amongst *Juniperus* sp. in the altitude of 3,000 - 4,000 m. **Map 2**.

Notes: This is named after Sir Henry Collett (1836-1901), a Military General in India during the British rule.

Key to the subsections of section Rhododendron

1.	Leaves very densely pilose beneath, the hairs completely obscuring the small golden scales	*Edgeworthia*
1a.	Leaves glabrous or variously hairy beneath, the hairs not obscuring the scales	2
2.	Style thin, declinate or straight, never markedly deflexed, usually longer than the generally declinate stamens	3
2a.	Style thick, markedly deflexed, atleast before anthesis, usually shorter than the actinomorphically arranged stamens..................................	17
3.	Lateral infloresences present; terminal inflorescences present or not	4
3a.	Lateral inflorescences usually absent ; all inflorescences terminal	6
4.	Corolla lobes shorter than the tube	5

4a.	Corolla lobes longer than the tube (rarely only slightly so)..................	7
5.	Corolla funnel-shaped, tube pubescent out side; calyx clearly lobed, lobes 2 - 3 mm.	*Virgata*
5a.	Corolla tubular to tubular-campanulate, pruinose and glabrous outside; calyx a mere rim or somewhat undulate	*Cinnabarina*
6.	Flowers pendulous; corolla fleshy, tubular to tubular-campanulate; nectaries present	*Cinnabarina*
6a.	Flowers neither pendulous nor fleshy; openly funnel-shaped and very zygomorphic; nectaries absent	*Triflora*
7.	Corolla lobes longer than the tube	8
7a.	Corolla lobes nearly as long as the tube or shorter	9
8.	Pedicel in line with the axis of the flower; corolla funnel-shaped, or almost hypocrateriform; scales undulate	*Lapponica*
8a.	Pedicel making an obtuse angle with the axis of the flower; corolla very openly funnel-shaped to almost rotate; scales crenulate or vesicular	*Saluenensia*
9.	Corolla 40 mm or more	*Maddenia*
9a.	Corolla 6 - 35 mm	10
10.	Corolla campanulate or tubular-campanulate, the sides of the tube parallel or nearly so	11
10a.	Corolla funnel-shaped, flaring from the base, the sides of the tube diverging	14
11.	Calyx reduced to a rim, sometimes slightly undulate, less than 2 mm; nectar copious, 5 large drops at the base of the corolla tube.................	*Cinnabarina*
11a.	Calyx conspicuously 5-lobed, the lobes more than 2 mm; nectar a mere sticky smear at the base of the ovary	12
12.	Seeds not winged, obscurely finned; inflorescences terminal, axillary inflorescences occasionally present	*Tephropepla*
12a.	Seeds winged and markedly finned ; inflorescences always terminal	13

13.	Inflorescence 1(-2)-flowered; calyx minute, densely lepidote ; stamens 10; ovary 5-celled; style glabrous, elepidote	*Monantha*
13a.	Inflorescence more than 3-flowered; calyx variable, ciliate; stamens more than 10 (-10 - 27); ovary 5 - 12-celled; style lepidote below	*Maddenia*
14.	Corolla pubescent outside ; small shrublets	*Uniflora*
14a.	Corolla lepidote outside ; tall shrubs	15
15.	Corolla white or yellow, drying greenish (*Boothia*) or yellow, tinged with red (*Trichoclada*) or purple to reddish-purple often with darker spots (*Baileya*) or white or cream, marked pink (*Camelliiflora*); style glabrous elepidote or lepidote	16
15a.	Corolla magenta to purplish with pink to dark red spots; style pubescent	*Lapponica*
16.	Corolla white or cream, marked pink; stamens 12 - 16 ; style glabrous, elepidote	*Camelliiflora*
16a.	Corolla yellow, tinged with red or purple to reddish purple, often with darker spots; stamens 5 - 10; style lepidote or glabrous.......................	18
17.	Leaf lower surface white-papillose; scales markedly dimorphic; larger brown, smaller golden..	*Glauca*
17a.	Leaf lower surface and scales various, not as above	19
18.	Scales vesicular	20
18a.	Scales not vesicular, clearly rimmed, the rims sometimes upturned	21
19.	Corolla pruinose, pink to purple	*Campylogyna*
19a.	Corolla not pruinose, yellow to white, sometimes flushed pink	*Lepidota*
20.	Style gradually tapering from the ovary, lepidote; pedicels shorter than the corollas..	*Boothia*
20a.	Style impressed into the ovary; glabrous; pedicels longer than the corollas	*Trichoclada*
21	Scales crenulate ; style impressed, glabrous	*Baileya*
21a.	Scales entire; ovary tapering into the lepidote style............................	*Boothia*

Subsection Baileya

Subsection **Baileya** Sleumer, Bot. Jahrb. 74:531 (1949). Cullen, Notes R.B.G. Edinb. 39(1) : 150(1980). Cox & Cox, The Encyclopedia of Rhododendron Species 234 (1997).

Syn.: Series *Lepidotum* sensu Hutchinson in Stevenson(ed.) The Species of Rhododendron 437 (1930) pro parte.

Type species : *R. baileyi* Balf.f.

A monotypic subsection.

R. baileyi Balf.f., Notes R.B.G. Edinb. 11:23 (1919). Cullen, Notes R.B.G. Edinb. 39(1) : 151(1980). Cox & Cox, The Encyclopedia of Rhododendron Species 234 – 235 (1997).

Syn.: *R. thyodocum* Balf.f. & Cooper, Notes R.B.G. Edinb. 11:148 (1919).

Shrub to 2 m high. Young branches straggly, lepidote. Leaves evergreen, 2 - 7 x 1.5 - 3 cm, narrowly elliptic, elliptic to obovate, apex obtuse or rounded, base cuneate, upper surface densely lepidote in young leaves, most scales deciduous on ageing, lower surface dark brown, densely lepidote, scales notched or toothed, overlapping. Inflorescence 5 - 9 flowered, sometimes up to 18. Calyx 5-lobed, lobes deltoid, 2 - 4 mm, lepidote, margins often setose. Corolla rotate, *ca* 3 cm across, scaly outside, reddish-purple or deep purple with or without darker spots. Stamens 10. Ovary lepidote; style impressed, sharply deflexed, shorter than the stamens, usually glabrous. Capsule small, lepidote.

Distribution and habitat: India : Sikkim. Bhutan, China (S. Xizang). In forests and open hill slopes, from 2,400 - 4,300 m. **Map 3.**

Notes: The species is named after Lt. Col. F.M. Bailey, a traveller in Tibet, China.

Subsection Boothia

Subsection **Boothia** (Hutchinson) Sleumer, Bot. Jahrb. 74:530 (1949). Cullen, Notes R.B.G.Edinb. 39 (1):133 (1980). Cox & Cox, The Encyclopedia of Rhododendron Species 235 (1997).

Syn.: Series *Boothii* sensu Hutchinson in Stevenson(ed.) The Species of Rhododendron 155(1930) pro parte; sensu Cowan & Davidian, Rhodo. Yearbook, 3:60 – 72 (1948) pro max. parte.

Small or medium sized shrubs, often epiphytic in the wild. Young growth often setose, setae persistent, sometimes glabrous (*R. leptocarpum)*. Leaves evergreen, glaucous or whitish-papillose beneath, scaly, scales rimmed or vesicular, deeply sunk in pits. Inflorescences terminal, 1 - 10-flowered; pedicels 1.5 - 5 cm long, rigid or flexuous. Calyx well-developed, lobed. Corolla broadly campanulate to rotate, yellow. Stamens 10, not declinate. Ovary densely lepidote, tapering into the style; style lepidote, sharply deflexed. Capsule lepidote. Seeds winged and finned.

Type species : *R. boothii* Nuttall

Key to the species of subsection Boothia

1.	Pedicels 2.5 - 5 cm long, slender, flexuous; calyx lobes reflexed at least in fruit; corolla 0.9 - 1.3 cm; young stems and leaf margins glabrous	*leptocarpum*
1a.	Pedicels not more than 1.5 cm long, stout, rigid; calyx lobes not reflexed; corolla 1.5 - 3 cm; young stems and leaf margins setose	2
2.	Leaves glabrous above except for a few setae on the midrib at the base and on the margins; scales vesicular; inflorescence 1 - 2 (- 3)- flowered	*megeratum*
2a.	Leaves above with a dense matted indumentum of twisted loriform setae; scales with rims; inflorescence more than 3-flowered	*boothii*

R. boothii Nuttall in Hooker's Kew Journ. 5:356 (1853). C.B. Clarke in Hook.f. Fl. Brit. India 3:470 (1882). Cullen, Notes R.B.G. Edinb. 39(1):134 (1980). Cox & Cox, The Encyclopedia of Rhododendron Species 236 (1997).

Syn.: *R. mishmiense* Hutchinson & Kingdon-Ward, Notes R.B.G. Edinb. 16:173 (1931).

Shrub up to 2 m, epiphytic or on rocks. Branchlets lepidote, indumentum densely matted with twisted bristles. Leaves 8 - 12.5 x 3.5 - 6 cm, ovate or ovate-oblong, thick and leathery, shortly acuminate at apex, upper surface with bristles on the midrib and margins, lower surface with dark brown scales their own diameter apart. Inflorescence 3 - 10-flowered; pedicels stout, up to 1.5 cm long, densely setose. Calyx lobes green, 7 - 15 mm, ovate to oblong, lepidote and ciliate. Corolla campanulate, 2 - 3 cm, pale to bright yellow, sometimes spotted, tube *ca* 1.5 cm, lepidote outside, pilose within. Ovary lepidote. Capsule ovoid, *ca* 1.5 cm long.

Distribution and habitat: India : Western & Eastern Arunachal Pradesh (Delei Valley in Mishmee Hills, Lohit district). China (S. Xizang). Epiphytic on forest trees or on mossy rocks in the altitudinal range of 1,800 - 3,000 m. **Map 3.**

Notes: The species is named after T.J. Booth, a plant collector in Bhutan in about 1850. Many of his specimens reported to be from Bhutan were actually collected in Arunachal Pradesh (Ludlow in Trans.Bot.Soc.Edinb.41:1972).

R. leptocarpum Nuttall ex C.B.Clarke in Hook.f., Fl. Brit. India 3:471 (1882).

Syn.: *R. micromeres* Tagg, Notes R.B.G. Edinb. 16:211 (1931). *R. pumilum* Nuttall in Hooker, Kew Journ. 5:354 (1853) non Hook.f.

Epiphytic or rarely free growing shrub up to 2 m, branches spreading. Bark dark mahogany-brown on mature stems, peeling. Young growth lepidote, glabrous. Leaves 3.8 x 1.5 - 3.6 cm, narrowly elliptic to elliptic, sometimes obovate, upper surface bright green, lower surface glaucous-grey and scaly, scales unequal, the smaller sunken in pits with rims upturned. Inflorescence 3 - 8 (- 13)-flowered ; pedicels slender, flexuous, lepidote, elongating in fruit up to 5 cm. Calyx lobed, lepidote, lobes 3 - 5 mm, spreading or reflexed. Corolla broadly campanulate, 0.9 - 1.4 cm long, yellow to white, tube 4 - 6 mm long, lepidote outside, pilose within. Stamens 10, filaments pilose in the lower part. Ovary lepidote. Capsule narrowly cylindric or often curved / sickle-shaped, lepidote, 1.2 - 1.5 cm long.

Distribution and habitat : India : Arunachal Pradesh. Eastern Bhutan, China (S.E. Xizang, N.W.Yunnan), N.E. Myanmar. Epiphytic on other larger rhododendrons and conifers or on mossy rock boulders in the altitudinal range of 2,400 - 4,300 m. **Map 3.**

Notes: The specific epithet derives its name from the thin (slender) fruit (*lepto* = thin, slender, *carpum* = capsule or fruit).

R. megeratum Balf.f. & Forrest, Notes R.B.G. Edinb. 12:140 (1920). Cullen, Notes R.B.G. Edinb. 39(1):136(1980). Cox & Cox, The Encyclopedia of Rhododendron Species 238(1997).

Syn.: *R. tapeinum* Balf.f. & Farrer, Notes R.B.G. Edinb. 12:164 (1920).

Shrub, 0.3 - 1 m, rarely epiphytic. Branchlets with persistent setae. Leaves 1.5 - 4 x 1 - 2 cm, elliptic, elliptic-obovate or orbicular, upper surface glabrous, midrib sparsely setose at the base, margins hairy, lower surface glaucous-whitish with vesicular scales sunken in pits; petiole bristly-hairy. Inflorescence 1 - 3-flowered, pedicels bristly-hairy, shorter than the flowers. Calyx lobes obovate, 7 - 10 mm, sparsely lepidote and ciliate outside. Corolla broadly-campanulate, 2 - 3 cm, yellow or cream with darker spots within, tube 8 - 10 mm. Stamens 10. Ovary lepidote; style short, sharply deflexed. Capsule ovoid-cylindric, 8 - 11 mm long.

Distribution and habitat: India : W. Arunachal Pradesh. China (S.E. Xizang, N.W. Yunnan), N. Mynamar. Rare in India, so far known only from W. Arunachal Pradesh (Tawang and Kameng districts bordering Xizang) in the altitudinal range of 2400 - 4100 m on mossy rocks and cliffs. **Map 3.**

Subsection Camelliiflora

Subsection **Camelliiflora** (Hutchinson) Sleumer, Bot. Jahrb. 74:533 (1949). Cullen, Notes R.B.G. Edinb. 39(1):138 (1980). Cox & Cox, The Encyclopedia of Rhododendron Species 239 (1997).

Syn.: Series *Camelliaeflorum* sense Hutchinson in Stevenson (ed.) The Species of Rhododendron 171 (1930).

A monotypic subsection.

Type species: *R. camelliiflorum* Hook.f.

R. camelliiflorum Hook.f., Rhododeondrons of Sikkim Himalaya t. 28 (1849). C.B. Clarke in Hook.f., Fl. Brit. India 3:470 (1882). Cox & Cox, The Encyclopedia of Rhododendron Species 239 (1997). **(Photo No. 19).**

Syn.: *R. sparsiflorum* Nuttall in Hooker's Kew Journ. 5:363 (1853). *R. cooperi* Balf.f., Notes R.B.G. Edinb. 10:91 (1917).

Shrub, up to 3 m, epiphytic or growing on rocks. Branchlets lepidote; bark reddish, peeling. Leaves 3.8 - 12.3 x 1.5 - 3.8 cm, oblong or narrowly elliptic, shortly acute, base obtuse or rounded, shining dark green, upper surface nearly glabrous,lower surface densely ferruginous scaly, scales contiguous, broadly rimmed. Inflorescence terminal, 1 - 3-flowered. Calyx lobes oblong, 5 - 11 mm, almost glabrous, fringed with scales. Corolla broadly tubular with spreading orbicular lobes, *ca* 3.8 cm across, white to deep rose, or rarely with a yellowish dash at the base, lepidote outside, villous within. Stamens 12 - 16, filaments pilose at base. Ovary 5 - 10-locular, lepidote; style usually shorter than stamens, sharply bent. Seeds conspicuously winged and finned.

Distribution and habitat: India : Sikkim, Western Arunachal Pradesh. E. Nepal, Bhutan. Frequent on forest trees, forest margins and cliffs in the altitudes of 2,700 - 3,700 m. **Map 3.**

Notes: The densely lepidote lower surface of the leaves, the small, somewhat flat, rotate 'camellia-like' flowers (from which it derives its specific epithet) with 12 - 16 stamens help easy identification of the species in the field.

Subsection Campylogyna

Subsection **Campylogyna** (Hutchinson) Sleumer, Bot. Jahrb. 74:531 (1949). Cullen, Notes R.B.G. Edinb. 39(1): 145(1980). Cox & Cox, The Encyclopedia of Rhododendron Species 240(1997).

Syn.: Series Campylogynum sensu Hutchinson in Stevenson (ed.) The Species of Rhododendron 182(1930); sensu Davidian in Rhodo. Yearbook 8:78(1954).

Small, usually prostrate shrublets. Branchlets lepidote, glabrous or sparsely pubescent. Leaves evergreen, small, papillose, with distant small vesicular deciduous scales. Inflorescence 1 - 3-flowered; pedicels long. Calyx 5-lobed. Corolla campanulate, pink to purple. Stamens 10, filaments pubescent at base. Ovary sparsely lepidote; style impressed, deflexed. Capsule lepidote; seeds obscurely finned.

Type species: *R. campylogynum* Franchet

A monotypic subsection.

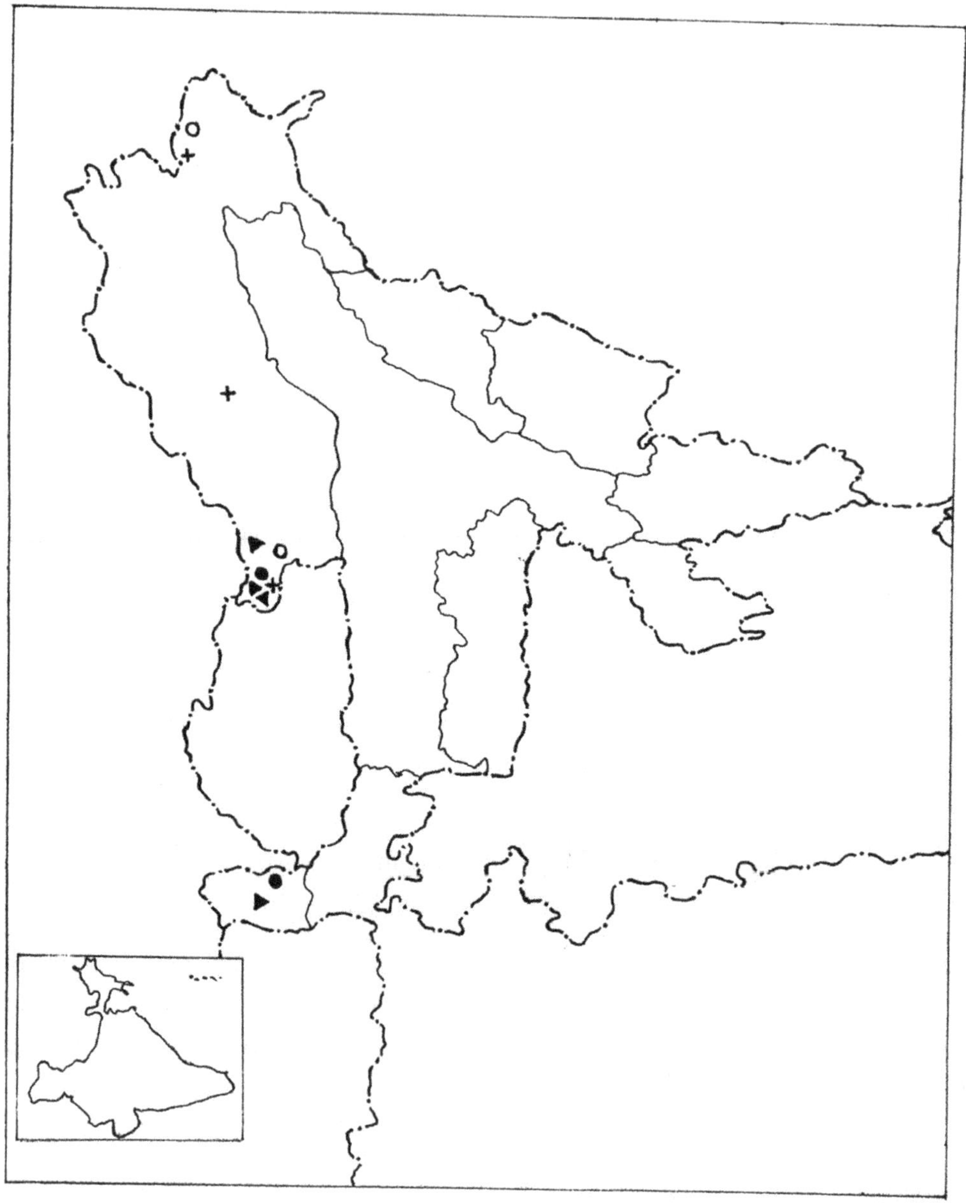

Map 3. ● R.baileyi. + R.leptocarpum. ▲ R.megeratum. O R.boothii. ▼ R.camelliflorum.

(Maps are only indicative and not to scale)

R. campylogynum Franchet, Bull. Soc. Bot. Fr. 32 : 10 (1885). Cullen, Notes R.B.G. Edinb. 39(1) : 145(1980). Cox & Cox, The Encyclopedia of Rhododendron Species 240(1997).

Syn.: R. caeruleo-glaucum Balf.f. & Forrest, Notes R.B.G. Edinb. 13: 34(1920).

Small, mound-forming shrublet. Branchlets lepidote. Leaves 0.7 - 3.7 x 0.3 - 1.2 cm, obovate to narrowly elliptic, upper surface dark green, shiny, lower surface glaucous or pale green, with deciduous, distant, vesicular scales. Inflorescence 1 - 3-flowered. Calyx 1 - 6 mm, usually lepidote. Corolla campanulate 1 - 2.3 cm long, very variable, pink, red or purple, sometimes cream-coloured, often with a bloom, usually held above the foliage on long pedicels. Ovary lepidote; style impressed on top of the ovary, sharply bent, glabrous. Capsule held on accrescent, long pedicel, sparsely lepidote.

Distribution and habitat: India : Eastern Arunachal Pradesh. China (S. & S.E. Xizang, Yunnan), N.E. Myanmar. Among larger rhododendrons, on rock cliffs, ledges and in moorland, in the altitudinal range of 2,400 - 4,900 m. **Map 4.**

Notes: The specific epithet *campylogynum* means with 'bent ovary'. The species is a preferred rock-garden plant.

Subsection Cinnabarina

Subsection **Cinnabarina** (Hutchinson) Sleumer, Bot. Jahrb. 74:534 (1949). Cullen, Notes R.B.G. Edinb. 39(1): 122(1980). Cox & Cox, The Encyclopedia of Rhododendron Species 243(1997).

Syn.: Subgenus *Keysia* Nuttall ex Hooker, Bot. Mag. 81:t. 4875 (1855).

Section *Keysia* (Nuttall) Maximowicz, Mem. Acad. Sci. St. Petersb. ser. 7,16(9):15 (1870).

Series *Cinnabarinum* sensu Hutchinson in Stevenson (ed.) The Species of Rhododendron 221 (1930).

Shrubs to 7 m, rarely epiphytic. Young growth lepidote. Leaves usually evergreen, often glaucous, variable in shape, lower surface densely lepidote, scales unequal, not contiguous, rimmed. Inflorescence terminal or axillary, flowers usually pendulous. Calyx small, disc-like or undulate, lepidote or ciliate. Corolla fleshy, waxy, tubular to campanulate, variable in colour, yellow, orange, red to plum-crimson or purple, nectar held in 5 drops in the corolla tube. Stamens 10, declinate, filaments pubescent below or rarely glabrous. Ovary lepidote, 5-locular; style impressed, straight, sometimes hairy near the base. Capsule lepidote. Seeds unwinged, obscurely finned.

Type species : *R. cinnabarinum* Hook.f.

Notes: The species in this subsection are easily identified by the fleshy, waxy, pendulous, bell-shaped flowers with five large droplets of nectar held in the nectaries at the base of the corolla tube.

Key to the species of subsection Cinnabarina

1.	Inflorescences all terminal; corolla 2.5 - 3.6 cm, tubular-campanulate; scales very narrowly rimmed	*cinnabarinum*
1a.	Inflorescences mostly lateral ; corolla up to 2 cm, narrowly tubular; scales broadly rimmed	*keysii*

R. cinnabarinum Hook.f., Rhododendrons of Sikkim Himalaya t. 8 (1849). Cullen, Notes R.B.G. Edinb. 39(1): 123(1980). Cox & Cox, The Encyclopedia of Rhododendron Species 244(1997).

Erect shrub, to 7 m high. Branches often purplish, lepidote. Bark shaggy, light brown on ageing, peeling. Leaves evergreen, 3 - 11.5 x 1.8 - 5.5 cm, broadly to narrowly elliptic to lanceolate, obtuse at apex, narrowed to cordate at base; upper surface green, usually elepidote, often glaucous, lower surface often glaucous, with more or less unequal, fleshy, narrowly rimmed scales, touching to their own diameter apart. Inflorescences all terminal, 2 - 9-flowered; pedicels lepidote. Calyx minute, 1 - 2 mm, undulate or disc-like, lepidote. Corolla fleshy, tubular to tubular-campanulate, semi-pendant, elepidote outside, usually with a waxy, pruinose bloom, 2.6 - 5 cm long, in shades of yellow (yellow-orange) to orange, pink, red, purple, plum-crimson or mixed colours. Stamens 10, filaments pubescent towards the base or rarely glabrous. Ovary densely lepidote; style puberulous at base. Capsule lepidote, *ca* 1 cm, cylindric.

Notes: This species is very variable and is divisible into three subspecies of which subsp. *cinnabarinum* and subsp. *xanthocodon* are reported from India.

Key to the subspecies of *R. cinnabarinum*

1.	Leaves relatively narrow, more than 2.5 times longer than their width, usually elepidote above; corolla more or less tubular-campanulate, red to orange coloured......................	subsp. *cinnabarinum*
1a.	Leaves relatively broader, less than 2.5 times longer than their width, persistently lepidote above; corolla usually campanulate, yellow coloured	subsp. *xanthocodon*

subsp. **cinnabarinum**

Syn. : *R. roylei* Hook.f., Rhododendrons of Sikkim Hiamalaya t. 7 (1849). *R. blandfordiiflorum* W.J. Hooker, Bot. Mag. 82 : t. 4930 (1856).

Leaves narrower, broadly oblanceolate, usually elepidote above. Corolla somewhat tubular-campanulate, red to orange to yellow or a combination of these colours.

The name *cinnabarinum* refers to its usually cinnabar-red coloured flowers.

Distribution and habitat: India : Sikkim; Darjeeling Hills, West Bengal. E. Nepal, Bhutan, China (S.E. Xizang). Common, often forming thickets as undergrowth of mixed forests and along hill slopes in the altitudes of 2100 - 4000 m (7000 - 13000 ft). Very elegant looking plants with shiny, dark green foliage and variously coloured pendent flowers. **Map 4.**

R.cinnabarinum Hook.f.

subsp. xanthocodon (Hutchinson) Cullen, Notes R.B.G. Edinb. 36:113 (1978). **(Photo No. 22).**

Syn.: *R. xanthocodon* Hutchinson, Gard. Chron. 95:409 (1934).

Leaves broader, elliptic or oblong-elliptic, persistently lepidote above. Corolla tubular-campanulate to campanulate, yellow, orange, apricot or purple.

xanthocodon means 'yellow bell' which refers to its yellow bell-shaped flowers.

Distribution and habitat : India : Arunachal Pradesh (Delei valley, Mishmi Hillls). Bhutan, China (S. Xizang). As undergrowth in temperate forests and forest margins, in the altitudes of 3050-3950 m and seems to be less frequent than subsp. *cinnabarinum.* **Map 4.**

R. keysii Nuttall in Hooker's Kew Journ. 55:353 (1853). Cullen, Notes R.B.G. Edinb. 39(1): 126(1980). Cox & Cox, The Encyclopedia of Rhododendron Species 246 (1997). **(Photo Nos. 23 & 24).**

Syn.: *R. igneum* Cowan, Notes R.B.G. Edinb. 19:235 (1937).

Straggling shrub, 1.2 - 6 m high, rarely epiphytic. Branchlets lepidote. Leaves 5 - 15 x 1.5 - 3.6 cm, elliptic to lanceolate, upper surface light to medium green, not glaucous, moderately lepidote. lower surface densely lepidote, scales unequal, flat, broadly rimmed. Inflorescence axillary and terminal, 2 - 6 flowers per bud often forming dense clusters; flowers pendulous. Calyx inconspicuous, undulate, lepidote, rarely ciliate. Corolla narrowly tubular, 1.5 - 2.5 cm long, deep red to salmon pink ; tube 1.5 - 2 cm long, lobes usually yellow. Stamens 10, filaments pubescent in the lower half. Ovary lepidote, slightly pubescent at the top; style pubescent at the base. Capsule cylindric, 1 cm, lepidote.

Distribution and habitat: India : Sikkim, Western Arunachal Pradesh (Kameng district). Bhutan and China (S. Xizang). Common in mixed and conifer forests and in thickets in the altitudinal range of 2400 - 3700 m. **Map 4.**

Named after a Mr. Keys.

Subsection Edgeworthia

Subsection **Edgeworthia** (Hutchinson) Sleumer, Bot. Jahrb. 74:532 (1949). Cullen, Notes R.B.G. Edinb. 39(1): 25(1980). Cox & Cox, The Encyclopedia of Rhododendron Species 247(1997).

Syn.: Series *Edgeworthii* sensu Hutchinson in Stevenson (ed.) The Species of Rhododendron 228 (1930).

Shrubs, up to 2.5 m. high, epiphytic or straggling among rocks. Leaves evergreen, often bullate on the upper surface, covered with dense indumentum of curled hairs, usually obscuring the yellow scales on the lower surface. Inflorescence terminal, 1 - 4 (- 6)-flowered. Calyx well-developed, conspicuously 5-lobed. Corolla very large or small, funnel-campanulate or campanulate, white, pale pink to yellow, fragrant or not. Stamens 10, filaments pubscent in the basal part. Ovary 5-celled, densely tomentose, style declinate, either longer or shorter than the stamens, or strongly deflexed from base. Capsule tomentose; seeds winged and finned.

Type species : *R. edgeworthii* Hook.f.

This subsection is characterized in having a dense indumentum of curled, brownish or whitish hairs which on the lower surface of the leaf obscure the small golden-yellow scales.

Key to the species of subsection Edgeworthia

1. Leaves strongly bullate and glabrous on the upper surface; corolla 35 - 75 mm.; style declinate, exceeding the declinate stamens .... *edgeworthii*

1a. Leaves smooth on the upper surface; corolla 15 - 22 mm.; style sharply deflexed downwards, shorter than the nearly actinomorphic stamens .... *pendulum*

R. edgeworthii Hook.f., Rhododendrons of Sikkim Himalaya t. 21 (1849). Cullen Notes R.B.G. Edinb. 39(1): 26(1980). Cox & Cox, The Encyclopedia of Rhododendron Species 247(1997). **(Photo No. 25).**

Syn.: *R. bullatum* Franch., Bull. Soc. Bot. Fr. 34:281 (1887).

Straggling shrub up to 3 m high, usually epiphytic or scrambling on rocks. Branchlets densely woolly tomentose. Leaves 6 - 15 x 2.5 - 5 cm., oblong-ovate, oblong-lanceolate or rarely elliptic; upper surface dark green, shiny, strongly bullate and glabrous; lower surface with small, distant golden scales completely obscured by the dense, woolly indumentum which is white when young, turning rusty-brown on maturity. Inflorescence 1 - 5-flowered; pedicels up to 2 cm long, densely tomentose. Flowers sweet scented and showy. Calyx large, 10 - 14 mm, deeply lobed, lobes oblong-orbicular, densely tomentose on the margins, scaly on the outer surface, finely pubscent on the inner surface. Corolla funnel-campanulate, 30 - 75 mm long, the tube (20-) 25 - 30 (-40) mm, white, sometimes flushed pink or rose, with or without a yellow blotch at the base, glabrous within, lepidote outside. Stamens 10, declinate, filaments densely pilose in the lower part.

Ovary densely tomentose; style long and straight, equalling the corolla, longer than stamens, tomentose and or sparsely scaly from above the base. Capsule densely tomentose, oblong-globose, *ca* 18x12 mm.

Distribution and habitat: India : Sikkim, W. Bengal, Arunachal Pradesh. Bhutan, East Myanmar, China (N., N.W. and Central Yunnan and S.E. Xizang). In dense mixed forest in the altitudes of 1800 - 4000 m; epiphytic or on rocks and cliffs. **Map 4.**

Notes: The species is named after M.P. Edgeworth of erstwhile Bengal Civil Service under the British rule. It is very easily recognized in the field by its bullate, dark green shiny leaves with dense cobwebby indumentum on the lower surface of leaves hiding the golden scales, and highly sweet scented large flowers.

R. pendulum Hook.f., Rhododendrons of Sikkim Himalaya t. 13 (1849). Cullen, Notes R.B.G. Edinb. 39(1): 27(1980). Cox & Cox, The Encyclopedia of Rhododendron Species 248(1997). **(Photo Nos. 26 & 27).**

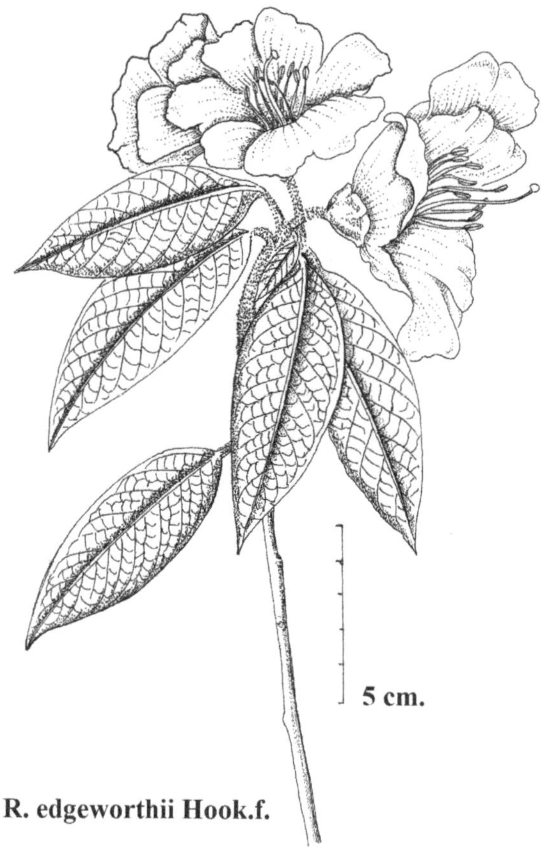

R. edgeworthii Hook.f.

Shrub, epiphytic or straggling, 0.3 - 1.2 m. Young shoots covered with densely woolly indumentum. Leaves 3 - 5 x 1.2 - 3 cm, oblong-elliptic, convex; upper surface smooth, shiny, without scales, lower surface with dense, whitish to light-fawn woolly indumentum. Inflorescence 2 - 3-flowered; pedicels densely tomentose. Calyx 5-lobed, lobes oblong or oblong-obovate, long ciliate, sparsely reddish-brown scaly and tomentose outside. Corolla openly funnel-campanulate, 1.5 - 2.2 cm, white or white flushed with pink or cream, tube 0.7 - 1 cm, scaly outside, glabrous within. Stamens 10, nearly actinomorphic, filaments pilose in the lower part. Ovary 5-locular, scaly and densely tomentose towards apex; style shorter than stamens, sharply deflexed, sparsely scaly at base. Capsule oblong, 1.2 cm.

Distribution and habitat: India : Sikkim. East Nepal, Bhutan, China (S. Xizang). In dense temperate forests in the altitudes of 2300 - 3700 m; epiphytic or amidst steep rock cliffs. **Map 4.**

Notes: The species epihthet *pendulum* indicates the hanging habit of the plant and its branches.

Subsection Glauca

Subsection **Glauca** (Hutchinson) Sleumer, Bot. Jahrb. 74:530 (1949). Cullen, Notes R.B.G. Edinb. 39(1):139(1980). Cox & Cox, The Encyclopedia of Rhododendron Species 251 (1997).

Syn.: Series *Glaucum* sensu Hutchinson in Stevenson (ed.) The Species of Rhododendron 294 (1930) pro parte.

Shrubs, to 2 m high. Bark frequently coppery, scaling. Leaves evergreen, 2 - 9 cm long, aromatic, white or greyish papillose beneath, scales dimorphic, the smaller white or yellow, the larger brown, stalked. Inflorescence 3 - 10-flowered. Calyx deeply 5-lobed. Corolla campanulate or tubular-campanulate, white, pink, purple. Stamens 10. Ovary lepidote, 5-locular; style impressed, glabrous, deflexed.

Type species: *R. glaucophyllum* Rehder

A distinctive subsection with aromatic leaves, glaucous undersurfaces with dimorphic golden and brown scales.

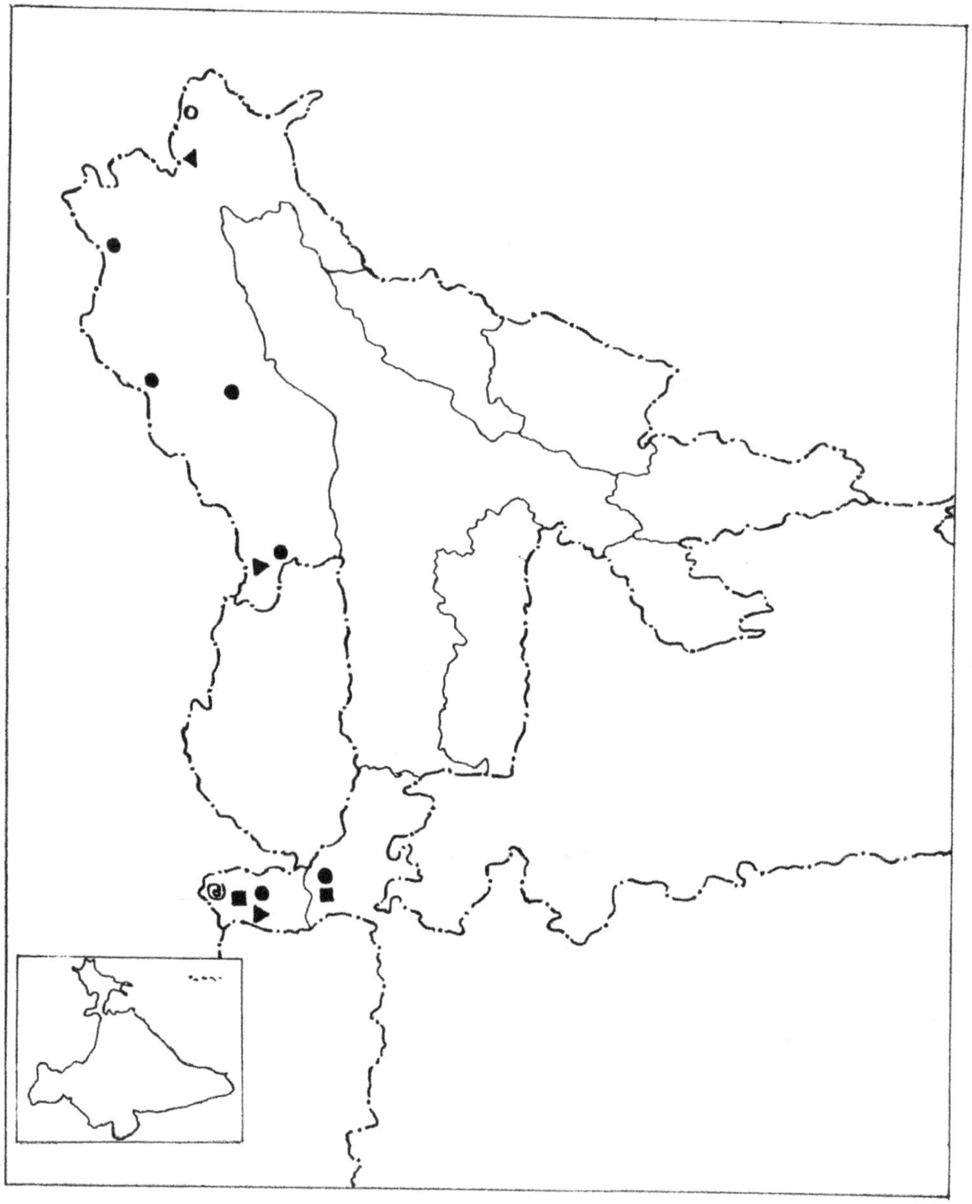

Map 4. ▲ R.campylogynum. ■ R.cinnabarinum ssp.cinnabarinum. O ssp.xanthocodon. ▼ R.keysii. ● R.edgeworthii. @ R.pendulum.

(Maps are only indicative and not to scale)

Key to the species of subsection Glauca

1. Calyx lobes acuminate, with a tuft of hairs inside at the tip; corolla lepidote outside, pinkish-purple, rose to pale pink....... *glaucophyllum*

1a. Calyx lobes rounded, without a tuft of hairs inside at the tip; corolla elepidote outside, dull-crimson to plum-purple.......... *pruniflorum*

R. glaucophyllum Rehder, Journ. Arn. Arb. 26:73 (1945). Cullen, Notes R.B.G. Edinb. 39(1) : 140(1980). Cox & Cox, The Encyclopedia of Rhododendron Species 253 (1997).

Shrub 1.5 m. Leaves very aromatic, 3.5 - 9.0 x 1.3 - 2.5 cm, lanceolate to oblong-lanceolate or elliptic-lanceolate, apex acute, base cuneate, upper surface dark brownish-green, lower surface glaucous to white and densely lepidote, scales dimorphic, the smaller pale yellow, the larger brownish. Inflorescence 4 - 10-flowered; pedicels lepidote, 13 - 20 mm long. Calyx large, foliaceous, ovate-acuminate, 6 - 9 (- 11) mm, lepidote at the base and along the margins, with a tuft of hairs inside at apex. Corolla campanulate or tubular-campanulate, 1.8 - 3 cm, pink or pinkish-purple, lepidote outside, tube 1.3 - 1.7 (- 1.9) cm long. Ovary 5-locular, lepidote; style impressed, sharply bent, shorter than the corolla or long and straight. Capsule ovoid, *ca* 1 cm long, lepidote.

Two varieties of this species are recognzed :

1. Corolla campanulate, pinkish; style shorter to corolla, sharply deflexed or bent .... var.*glaucophyllum*

1a. Corolla tubular-campanulate, rose pink; style longer than corolla, straight or declinate .... var. *tubiforme*

var. **glaucophyllum**

Syn.: *R. glaucum* Hook.f., Rhododendrons of Sikkim Himalaya t. 17 (1849) non Sweet, Hort. Brit. ed. 2, 344 (1830).

Distribution and habitat: India : Sikkim. Nepal, Bhutan, China (S. Xizang). Along rocky slopes and forest clearings in the altitudinal range of 2700 - 3700 m. **Map 5.**

var. **tubiforme** Cowan & Davidian, Rhodo. Yearbook 3:86 (1948).

Distribution and habitat: India : Kameng district, Western Arunachal Pradesh. E. Bhutan, China (S.E. Xizang). In forests and on rocks from 2,700 - 3,700 m altitude. **Map 5.**

Notes: The species epithet *glaucophyllum* refers to its bluish-grey leaves.

R. pruniflorum Hutchinson in Stevenson (ed.) The Species of Rhododendron 302 (1930). Cullen, Notes R.B.G. Edinb. 39(1):143(1980). Cox & Cox, The Encyclopedia of Rhododendron Species 255 (1997).

Syn.: *R. tsangpoense* var. *pruniflorum* (Hutchinson) Cowan & Davidian, Rhodo. Yearbook 3:90 (1948). *R. sordidum* Hutchinson, Rhodo. Soc. Notes 3: 286 (1932).

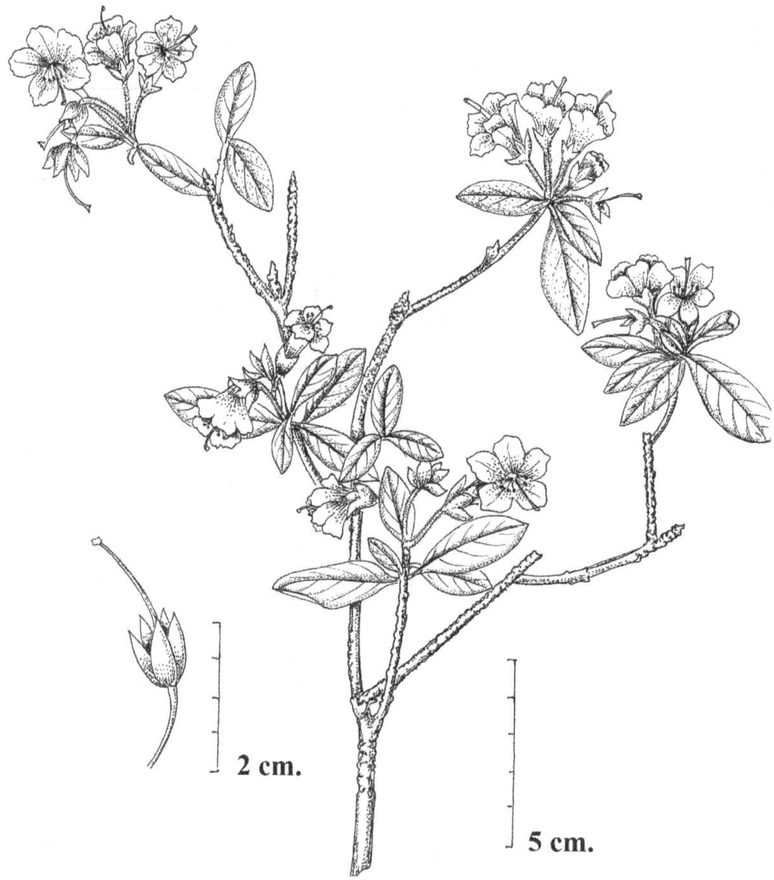

R.glacophyllum Rehder

Bushy shrub, to 1 m high. Bark brown, shredding. Leaves aromatic, 3 - 4.2 x 1.4 - 2.5 cm, ovate to oblong-elliptic, rounded at both ends,upper surface dark green, more or less elepidote, lower surface very glaucous and densely scaly, scales dimorphic, the smaller nearly contiguous (touching), pale yellow, the larger brown. Inflorescence 3 - 7-flowered; pedicels up to 3 cm, lepidote. Calyx prominent, lobes to 0.5 cm, rounded. Corolla campanulate, 1 - 1.3 cm long, dull crimson to plum purple, elepidote outside, tube 5-8 mm long. Stamens pubescent nearly their entire length. Capsule ovoid, 4 - 6 mm.

Distribution and habitat: India : Eastern Arunachal Pradesh. N. Burma, China (S.E. Xizang). In forests, rhododendron thickets and along rocky hill slopes. **Map 5.**

Notes: *Pruniflorum* means with flowers like those of *Prunus* (Plum).

Subsection Lapponica

Subsection **Lapponica** (Balf. f.) Sleumer, Bot. Jahrb. 74:535 (1949). Cullen, Notes R.B.G. Edinb. 39(1): 92(1980). Cox & Cox, The Encyclopedia of Rhododendron Species 259(1997).

Syn.: Series *Lapponicum* Balf.f., Notes R.B.G. Edinb. 9:298 (1916).

Small shrubs. Leaves evergreen, more or less aromatic, usually papillose beneath, 0.2 - 7 x 0.1 - 2.6 cm, densely lepidote with early deciduous leaf-bud scales, scales with broad, undulate margins. Inflorescence terminal, 1 – many-flowered; pedicels short and in line with the axis of the flower. Calyx usually distinctly 5-lobed, scaly and ciliate along the margins. Corolla broadly funnel-shaped, deep purple, pinkish, magenta or violet coloured. Stamens 5 - 10 (- 11). Ovary 5-celled, lepidote; style straight or declinate. Seeds without wings, obscurely finned.

Type species : *R. lapponicum* (L.) Wahlenberg

This large subsection with 27 species (Cullen, Notes R.B.G. Edinb. 39(1): 93-95, 1980) is represented in India by only two species, namely *R. nivale* and *R. setosum.*

Key to the species of subsection Lapponica

1. Erect shrub up to 1.2 m tall, branchlets densely bristly; inflorescence 3 - 8-flowered ; corolla bright purplish pink or wine red coloured ; calyx reddish or crimson-purple, lepidote or loriform-ciliate .... *setosum*

1a. Prostrate cushion-like shrub, under 1m high, branches and branchlets densely scaly; inflorescence 1 (- 2)-flowered; corolla purplish to lilac or violet coloured; calyx dark scaly outside, or ciliate on margins .... *nivale* subsp. *nivale*

R. nivale Hook.f. subsp. **nivale** Rhododendrons of Sikkim Himalaya 29(1849). Cullen, Notes R.B.G. Edinb. 39(1): 105(1980). Cox & Cox, The Encyclopedia of Rhododendron Species 270(1997).

Syn.: *R. paludosum* Hutchinson & Kingdon-Ward, Notes R.B.G. Edinb. 16:175 (1931).

Prostrate, much branched, cushion-like shrub, attaining 1 m high. Branches densely scaly. Leaves 0.5 - 0.9 x 0.2 - 0.5 cm, usually elliptic or broadly elliptic, ovate or rounded, apex obtuse, rounded or shortly acute, with or without a minute mucro, base cuneate, lower surface yellowish to fawn, densely scaly, scales dimorphic, majority pale gold, with few to many darker ones, widely spaced regularly. Inflorescence 1 - 2-flowered (rarely 3-flowered); pedicels 0.5 - 1.5 mm, scaly, sometimes also pubescent. Calyx obsolete, or with 2 - 4 mm long lobes, scaly and sometimes pubescent at the base. Corolla 0.9 - 1.3 cm long, broadly funnel-shaped, bright purplish pink or wine red, pubescent in the throat, with or without scales outside. Stamens usually 10, filaments pubescent in the lower part, longer or shorter than the corolla. Ovary lepidote; style variable in length, usually longer than stamens.

Distribution and habitat: India : Sikkim. Nepal, Bhutan, China (S. & S.E.Xizang). Frequent along open mountain slopes in the altitudes ranging from 3000 - 5800 m. The alpine forms exhibit shortest growth rate and smallest leaves. This subspecies is the *Rhododendron* recorded from the highest ever altitude, growing in the snow-line areas at 5800 m in the Himalyas and is aptly named *nivale* meaning 'plant of snowy areas'. **Map 5.**

R. setosum D. Don, Mem. Wern. Soc. 3:409 (1821). Cullen, Notes R.B.G. Edinb. 39(1): 109(1980). Cox & Cox, The Encyclopedia of Rhododendron Species 276(1997).

Bushy shrub, up to 1.5 m tall, often smaller. Branchlets densely clothed with stiff bristly hairs and scales. Leaves 1 - 1.9 x 0.6 - 0.8 cm, elliptic, oblong or obovate, apex rounded and obtuse with a short mucro, nearly rounded at base, margins hairy, upper surface dark green with persistent, yellow, more or less vesicular scales, sometimes with loriform-setae, lower surface with dimorphic scales, scales vesicular and golden-yellow, and flat and pale brown; petiole bristly. Inflorescence 1 - 6-flowered, rarely up to 8-flowered; pedicels lepidote and filliform-acicular pubescent. Calyx 5 - 8 mm, reddish or crimson- purple, lepidote or occasionally loriform-ciliate. Corolla 1.5 - 1.8 cm long, openly funnel-shaped, purple or pinkish to wine red, tube 7 - 8 mm, pilose inside, glabrous outside. Stamens 10, exerted, filaments pubescent in the lower half (towards the base). Ovary lepidote, filiform-acicular pubescent towards the apex; style longer than the stamens, glabrous. Capsule lepidote, to 5 mm long, oblong.

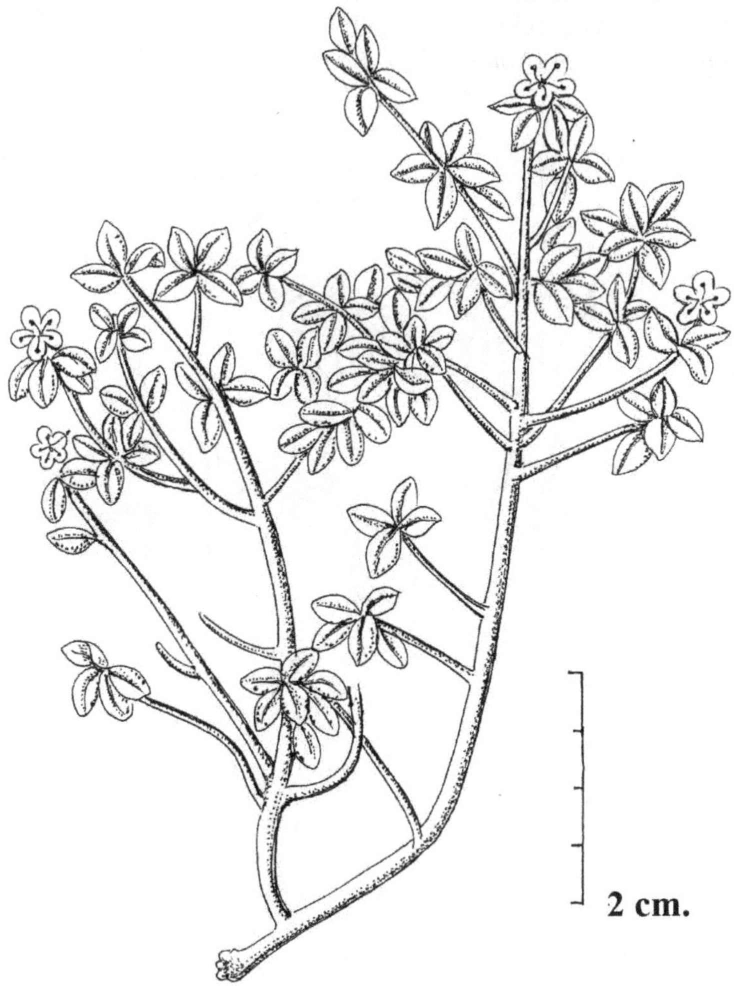

R. nivale Hook.f.

Distribution and habitat: India : Sikkim and Darjeeling Hills in West Bengal. Bhutan, China (S. Xizang-Chumbi valley). Frequent in open hill sides and slopes in the altitudinal range of 2700 - 4000 m. **Map 5.**

Notes: A distinctive species both in flower and foliage. This is the only species with bristles (setae after which its name *setosum*) on the branches, leaves and lower surface of leaves in the subsection *Lapponica*, because of which Philipson & Philipson (*Notes R.B.G. Edinb.* 34:1-71, 1975) transferred it under a separate monotypic subsection *Setosa*, with which Cox & Cox (*The Encyclopedia of Rhododendron Species* 277, 1997) are in agreement. However, this species is retained under the subsection *Lapponica* following Cullen's taxonomic treatment (*Notes R.B.G. Edinb.* 39(1) : 93, 1980).

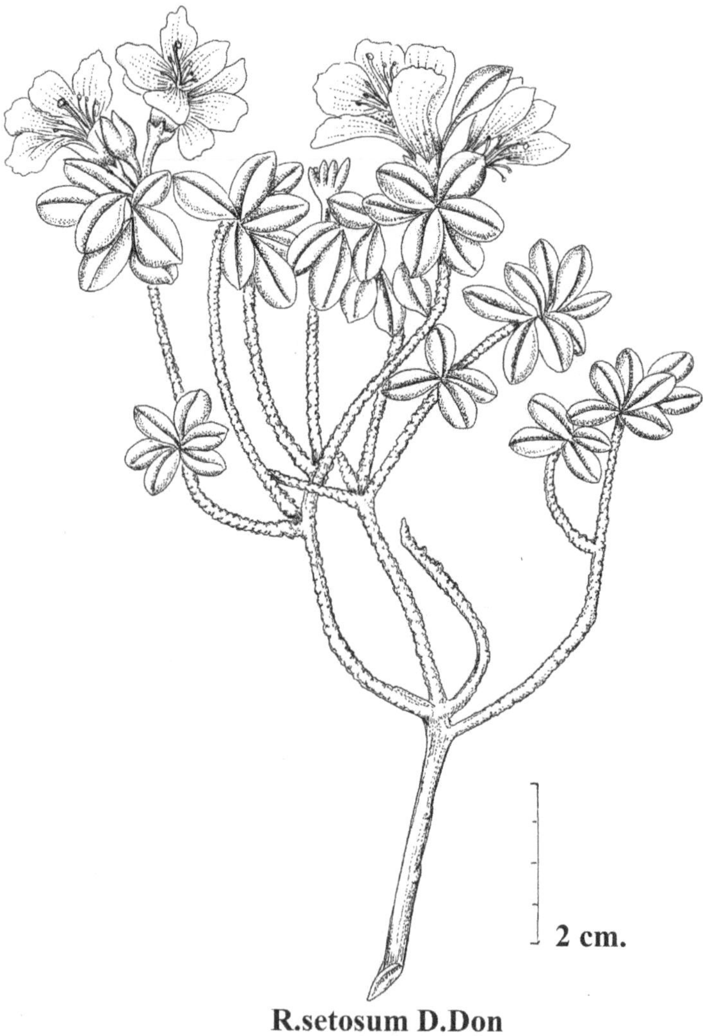

R.setosum D.Don

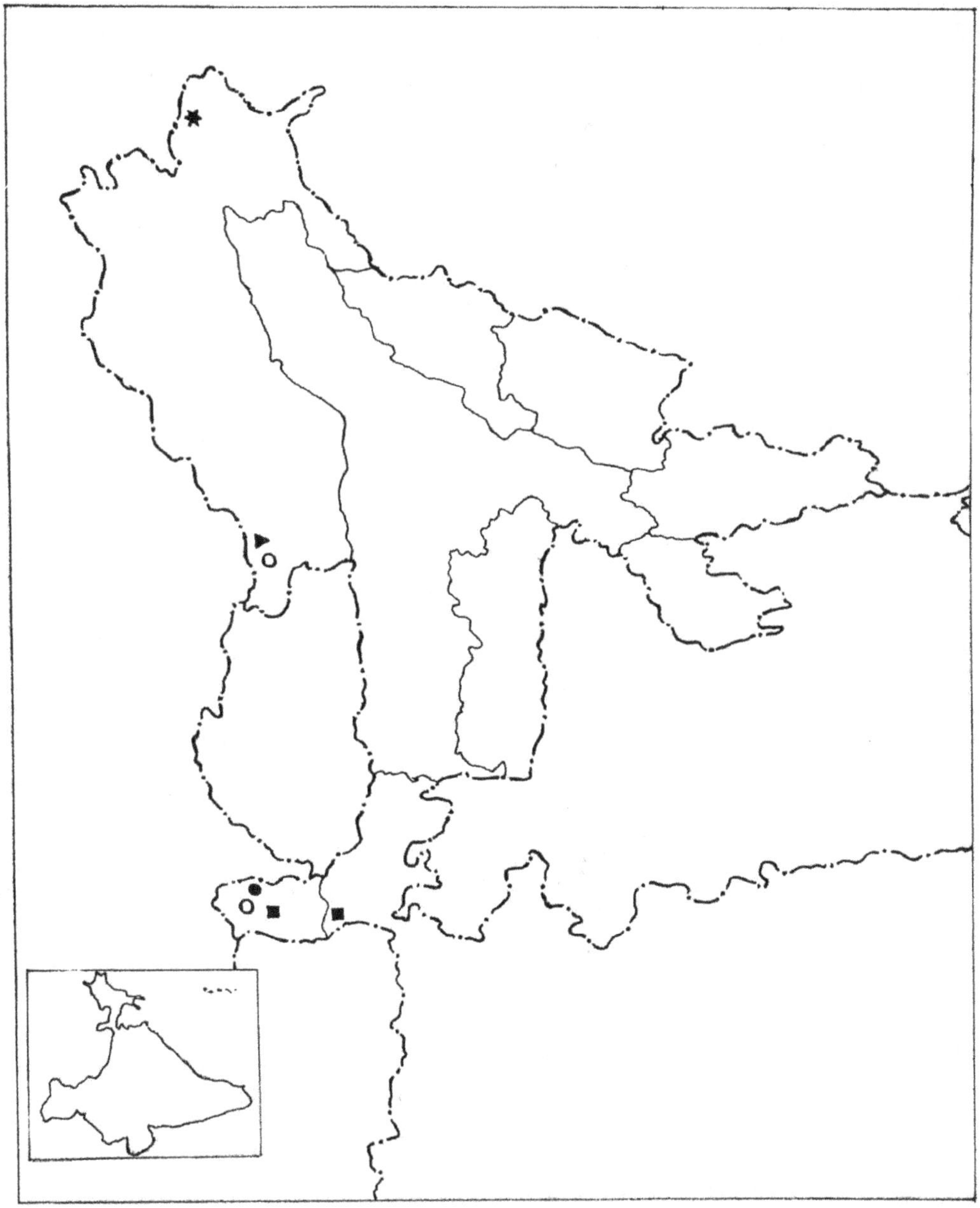

Map 5. ○ R.glaucophyllum var.glaucophyllum. ▼ var.tubiforme. ✱ R.pruniflorum. ■ R.setosum. ● R.nivale ssp.nivale.

(Maps are only indicative and not to scale)

Subsection Lepidota

Subsection **Lepidota** (Hutchinson) Sleumer, Bot. Jahrb. 74:531 (1949). C.B. Clarke in Hook.f., Fl. Brit. India 3:471 (1882). Cullen, Notes R.B.G. Edinb. 39(1) : 148(1980). Cox & Cox, The Encyclopedia of Rhododendron Species 284(1997).

Syn.: Series *Lepidotum* sensu Hutchinson in Stevenson (ed.) The Species of Rhododendron 437(1930) pro parte.

Type species: *R. lepidotum* Wallich ex G. Don

In India this subsection is represented by *R. lepidotum* and a recently described species, namely, *R. nayarii.*

Key to the species of subsection Lepidota

1. Terrestrial shrubs, 1.5 - 2 m high, often stoloniferous; young branches densely lepidote; inflorescence usually 2 - 3-flowered; pedicels 1.5 - 2.5 cm long, densely lepidote .... *lepidotum*

1a. Epiphytic shrublets, under 0.5 m high; young branches densely brownish pilose, older branches glabrescent; inflorescence 1-flowered; pedicels 0.5 - 1 cm long, densely pilose and lepidote .... *nayarii*

R. lepidotum Wallich ex G. Don, Gen. Hist. Dichlam. Pl. 3:845 (1834). C.B. Clarke in Hook.f., Fl. Brit. India 3:471 (1882). Cullen, Notes R.B.G. Edinb. 39(1):149(1980). Cox & Cox, The Encyclopedia of Rhododendron Species 284(1997).

Syn.: *R. elaeagnoides* Hooker, Rhododendrons of Sikkim Himalaya t. 23 (1849). *R. obovatum* Hooker, op. cit. consp. 6. *R. salignum* Hooker, op. cit. t. 23A.

Evergreen shrub, to 2 m tall, often stoloniferous. Young branches densely lepidote. Leaves 0.4 - 3.8 x 0.4 - 1.2 cm, narrowly elliptic, obovate or lanceolate, thick, upper surface dark green and densely scaly, lower surface pale greyish-green and densely covered with large, overlapping, brownish scales with translucent rims. Inflorescence 1 - 3 (rarely – 4)-flowered; pedicels lepidote, 1.2 - 2.5 cm long. Calyx variable in shape and colour, lepidote. Corolla rotate-campanulate, 1.2 - 2.5 cm in diameter, white, yellow, pink, red or various shades of purple, densely scaly outside. Ovary scaly; style very short, deflexed. Capsule cylindric, 4 - 6 mm, scaly.

Distribution and habitat : India, throughout the Himalya from Kashmir to Arunachal Pradesh. Nepal, Bhutan, N.E. Burma, China (N.W. Yunnan & S. & S.E. Xizang). In varied habitats starting from open forests, mountain slopes and moorlands in the altitudes ranging from 2,400 - 4,900 m. **Map 6.**

Notes: *Lepidotum* means clothed with scales.

R. nayarii G.D. Pal in Journ. Bombay Nat. Hist. Soc. 98 : 258 (2001).

Epiphytic branched shrublet, branches reaching up to 50 cm, slender; young shoots densely brownish-piloe becoming scabrid with age. Leaf-buds conical to ovoid, perulae imbricate, finely ciliate. Leaves pseudowhorled at apices of branchlets, elliptic to elliptic-oblong, rarely obovate, thick,10 - 20 x 6 - 12 mm, acute or rounded at base, obtuse or rounded at apex with a short mucro, margins recurved with loriform setae, usually densely scaly on both the surfaces; scales disc-like, sparse, golden-brown, sunken in pits, umbrella-like; petioles 1 - 7 mm, pilose and scaly. Flowers solitary, terminal, pedicels 5 - 10 mm long, pilose and lepidote. Calyx deeply 5-lobed, lobes 5 - 10 x 4 - 6 mm, oblong-spathulate, sparsely lepidote, margins acicular. Corolla campanulate to funnel-campanulate, yellow, tube 6 - 10 mm long, lepidote outside, lobes sub-orbicular, sparsely lepidote. Stamens 10, filaments pilose in the middle. Ovary 5-celled, densely lepidote; style deflexed, 3 - 4 mm long, glabrous, stigma 5-lobed.

Distribution and Habitat : India : Arunachal Pradesh. Endemic. In moist shady forest on large tree trunks and branches amidst ferns, mosses etc. in the altitude of about 3000 m. **Map 6.**

Notes: This is a recently described species collected from Talley valley in Lower Subansiri district, Arunachal Pradesh in April 1989 and is perhaps known only from the Type collections. Closely allied to *R. lowndesii* which is endemic to Nepal, this species is named in honour of Dr. M.P. Nayar, a former Director of the Botanical Survey of India.

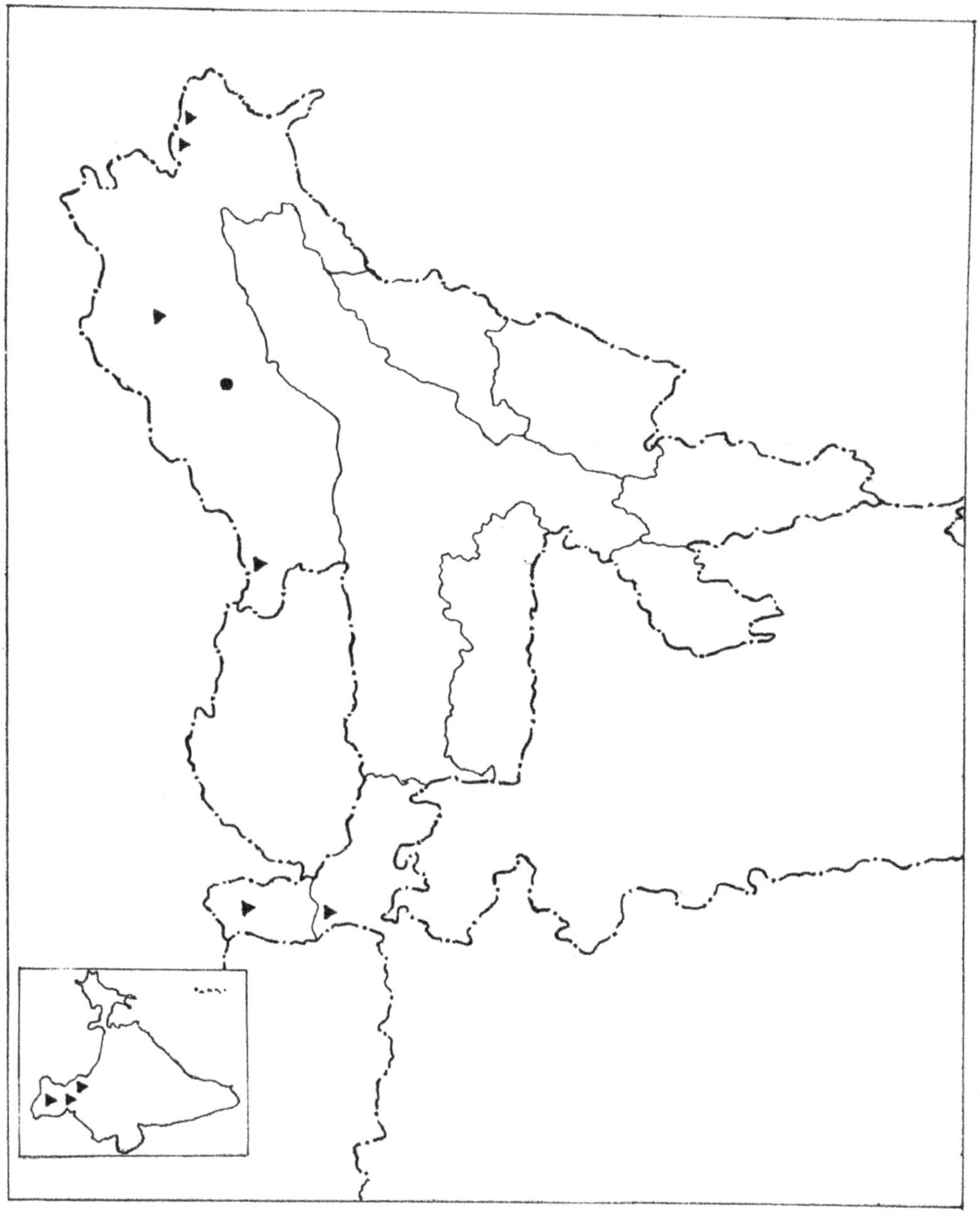

Map 6. ▼ R.lepidotum. ● R.nayarii.

(Maps are only indicative and not to scale)

Subsection Maddenia

Subsection **Maddenia** (Hutchinson) Sleumer, Bot. Jahrb. 74:533 (1949). Cullen, Notes R.B.G. Edinb. 39(1): 29(1980). Cox & Cox, The Encyclopedia of Rhododendron Species 286(1997).

Syn.: Series *Maddenii* sensu Hutchinson, Notes R.B.G. Edinb. 12:1 (1919); in Stevenson(ed.)The Species of Rhododendron 447 (1930).

Shrubs or small trees up to 15 m high, often epiphytic. Young branches lepidote, often loriform-setose. Leaves evergreen, lepidote, soon becoming elepidote, lower surface whitish or greyish, scaly, scales unequal, variously dense. Inflorescence terminal, 1 - 2-flowered, flowers usually fragrant. Calyx variable, mere rim-like or large, 5-lobed, ciliate. Corolla funnel-campanulate, rarely campanulate or very narrowly funnel-campanulate; corolla tube usually 5-grooved, white, pink or yellow. Stamens 8 - 27, filaments usually pilose towards the base. Ovary 5 - 12-locular, lepidote; style impressed or tapering from top of the ovary, lepidote at the base, declinate. Capsule lepidote; seeds winged and finned.

Type species: *R. maddenii* Hook.f.

Distribution and habitat: India : Eastern Himalaya : Sikkim, Arunachal Pradesh. Nepal, Bhutan, South Myanmar, W.& S. Yunnan, S.W. China, Thailand, Laos, Vietnam.

Notes: The subsection *Maddenia* consists of the most largest-leaved and largest-flowered lepidote species in the subgenus *Rhododendron* (excluding the Section *Vireya*). Many species are epiphytic and straggly shrubs usually with white and / or pale pink or yellow flowers; white flowers are most fragrant while the yellow ones have little or no scent.

Cox and Cox (*The Encyclopedia* of *Rhododendron Species* 286, 1997) state that taxonomically the species in subsection *Maddenia* to be amongst the most confusing in the genus.

Key to the species of subsection Maddenia

1.	Stamens (15 -) 17 - 25 (- 27); ovary 10 - 12-locular	*maddenii*
1a.	Staments 10 - 12, rarely 15; ovary 5 - 7-locular	2
2.	Pedicels and calyx whitish-pruinose, pedicels divaricate-recurved in fruit; capsule not exceeding the persistent calyx	*megacalyx*
2a.	Pedicels and calyx not whitish-pruinose, pedicels erect or spreading in fruit; capsule clearly exceeding the persistent calyx	3

3.	Leaf with the main vein raised on the upper surface; calyx deeply lobed, not ciliate or ciliate with filiform-acicular hairs	4
3a.	Leaf with the main vein totally impressed on the upper surface; calyx fringed with loriform setae	7
4.	Leaves with the secondary and tertiary veins forming a prominent reticulum on the lower surface; scales unequal, the smaller rimless, the larger with broad ascending irregular rims	*nuttallii*
4a.	Leaves with secondary and tertiary veins not forming a reticulum on the lower surface or obscurely so near the margins; scales nearly equal, not as above	5
5.	Pedicels pubescent as well as lepidote; calyx lobes pubescent outside......	*dalhousiae*
5a.	Pedicels not pubescent, lepidote only; calyx lobes ciliate or margined with scales	6
6.	Calyx lobes ciliate, cilia filiform-acicular	*lindleyi*
6a.	Calyx lobes not ciliate, margined with scales	*taggianum*
7.	Style arising out of a conspicuous depression on top of the ovary	8
7a.	Style not as above, ovary gradually tapering into the style	11
8.	Calyx more than 5 mm, lobes nearly equal, herbaceous; corolla up to 5 cm; style elepidote	*ciliatum*
8a.	Calyx less than 4 mm, if longer the lobes unequal, not herbaceous; corolla usually longer than 5 cm; style lepidote towards base	9
9.	Calyx disc-like, not hairy; corolla 10 cm long; epiphytic	*dendricola*
9a.	Calyx disc-like or rim-like with hairs; corolla not more than 7.5 cm long; terrestrial or epiphytic	10
10.	Corolla tubular funnel-shaped, 7 cm long ; calyx rim-like, hairy	*coxianum*
10a.	Corolla openly funnel-campanulate, 4 - 7.5 cm long; calyx disc-like, lepidote, with a few hairs	*formosum*
11.	Calyx disc-like, lepidote, margins ciliate	*veitchianum*
11a.	Calyx short, not disc-like and not lepidote, hairy	12
12.	Corolla funnel-campanulate, 4.6 - 6 cm long, not pubescent, yellowish-blotched at base	*johnstoneanum*
12a.	Corolla funnel-shaped, 6 - 7 cm long, pubescent all over or densely pilose, greenish blotched at base	*walongense*

Notes: Taxonomically, the species in the subsection *Maddenia* are amongst the most variable and confusing in the genus. J. Hutchinson (1919-1930) revised the Series *Maddenii* following Balfour's classification of the genus. He relied much on characters such as density of scales on the corolla and leaves, leaf shapes, leaf bases etc., which were found to be highly inconsistent by H. Sleumer(1949 and 1958) when specimens other than the Types were examined. Cox and Cox (*The Encyclopaedia of Rhododendron Species* 286, 1997) state that much of the cultivated stock does not match their wild plants and that many so-called species were named from cultivated specimens. They further remark about the great scarcity of herbarium collections for many of the so-called species, several of which are only known from their Type specimens.

The system of grouping outlined by Cox and Cox (1997) is followed in this work which roughly corresponds to the old classification of subseries as given below.

Group 1: Maddenii Alliance

Group 2: Dalhousiae Alliance

Group 3: Megacalyx

Group 4: Cilicalyx Alliance

4(a). R. ciliatum, valentinianum, formosum etc.

4(b). Johnstoneanum Aggregate

4(c). Cilicalyx Aggregate

Group 1- Maddenii Alliance

Now considered a single variable species with stamens (15 -) 17 - 25 (- 27) and ovary 8-12-locular.

R. maddenii Hook.f., Rhododendrons of Sikkim Himalaya t.18 (1849). Cullen, Notes R.B.G. Edinb. 39(1): 33(1980). Cox & Cox, op.cit.

Shrub, up to 6 m high, branches often sprawling or straggling, sometimes trailing on cliffs and steep slopes, rarely epiphytic. Bark rough and flaky. Leaves 5 - 20 x 2 - 8 cm, lanceolate, oblong-lanceolate, broadly ovate or elliptic, acute or obtuse at apex, cuneate at base, upper surface elepidote, lower surface densely lepidote, scales contiguous to overlapping, variable in size and shape. Inflorescence 2 - 11-flowered; pedicels 5 - 15 (- 20) mm, lepidote. Calyx deeply 5-lobed, lobes oblong to oblong-lanceolate, glabrous, 2.5 - 20 mm, margins erose. Corolla tubular-funnel-shaped, 4.5 - 12.5 cm long and to

12.5 cm wide, white, white flushed with pink or purple, rarely wholly pink, usually with a yellow blotch at the base, sweet scented; corolla tube 3 - 6 cm long, densely lepidote outside. Stamens (15 -) 17 - 25 (- 27), filaments glabrous or sparsely pilose at base. Ovary (8 -) 10 (- 12)-locular, densely lepidote, tapering into the style; style lepidote over most of its length. Capsule 10 - 20 mm long, ovoid-globose to oblong, 10 - 12-ridged, lepidote.

Distribution and habitat: India : Sikkim, Western Arunachal Pradesh, Manipur. In a variety of habitats, in dense forests, on open ridges and slopes, amidst rocks and cliffs, in the altitudes between 1500 - 3700 m.

Notes: *R. maddenii* is easily identified from other species in the subsection in having many stamens (17 - 25) and 8 - 12-celled ovary while other species have only stamens up to 12 and 5 - 7-celled ovary.

The species is named after Lt. Col. E. Madden, a traveller in India.

A complex and highly variable species representing two highly integrading subspecies.

1. Leaves less than 6 cm broad, usually 6 - 11 x 2.8 - 6 cm, often obovate; stamens often glabrous; capsules ovoid-globose, rounded to the apex .... subsp. *maddenii*

1a. Leaves more than 5.5 cm broad, usually 9 - 18 x 5.5 - 8 cm, usually more or less elliptic; stamens usually pubescent; capsule oblong-cylindrical, abruptly rounded to almost truncate at apex .... subsp. *crassum*

subsp. **crassum** (Franchet) Cullen, Notes R.B.G. Edinb. 36:107 (1978). Fig. 3c, d, p.19.

Syn. *R. crassum* Franchet, Bull. Soc. Bot. Fr. 34:282 (1887). *R. manipurense* Balf.f. & Watt, Notes R.B.G. Edinb. 10:119 (1917). *R. odoriferum* Hutchinson, Gard. Chron. 82:30 – 32 (1927).

Distribution and habitat: India : Manipur. Burma, China (N., N.W., W. & S.W. Yunnan, S.E. Xizang), Vietnam. **Map 7.**

Notes: This is the more easterly occurring and hardier subspecies. Plants under *Manipurense* Group have larger than average leaves while *Odoriferum* Group have small leaves and tubular flowers (Cox and Cox, 1997).

subsp. **maddenii. (Photo Nos. 30-32).**

Syn. R. brevitubum Balf. f. & R.E. Cooper, Notes R.B.G. Edinb.12:24 (1919), non J.J. Smith. *R. brachysiphon* Balf.f. ex Hutchinson, Notes R.B.G. Edinb. 12:24 (1919). *R. jenkinsii* Nutt. in Hooker's Kew Journ. 5:361 (1853).

Distribution and habitat: India : Sikkim, Western Arunachal Pradesh. Bhutan. **Map 7.**

Notes: This is the more westerly form of the species and is usually less hardier than subspecies *crassum*. Plants under *Brachysiphon* Group (Bhutan) often have pink flowers while certain other forms of *Polyandrum* Group (Bhutan) have the largest flowers (Cox & Cox, 1997).

Group 2. Dalhousiae Alliance

This alliance is characterized by the leaves with main vein raised on the upper surface; pedicels and calyx not covered with waxy, whitish 'bloom'; calyx large, conspicuous, deeply lobed, not ciliate or ciliate with filiform-acicular hairs; stamens 10 - 15; ovary 5-celled; pedicels erect to spreading in fruit; capsule longer than the persistent calyx; ovary tapered into the style.

R. dalhousiae Hook.f., Rhododendrons of Sikkim Himalaya t. 1 non t. 2 (1849). Cullen, Notes R.B.G. Edinb. 39(1):37(1980). Cox & Cox, The Encyclopedia of Rhododendron Species 289(1997).

Epiphytic or sub-erect sprawling shrub, up to 3 m high. Young shoots covered with small bristles. Bark brown, cinnamon, peeling. Leaves only at ends of branches, 6 - 20 x 2.5 - 7.5 cm, obovate to oblanceolate, tapered to the base, more or less rounded at apex, lower surface greyish or brownish-green with small reddish scales more than their own diameter apart, leaf margins often crenulate or bristly. Inflorescence 2 - 6-flowered; pedicels 1.5 - 2 cm, lepidote and pubescent, accrescent in fruit. Calyx conspicuous, deeply 5-lobed, 10 - 15 x 5 - 10 mm, lepidote at base, nearly glabrous or sparsely hairy in the middle. Corolla tubular-campanulate, white or creamy yellow, sometimes with 5 red bands running from the base of the tube to the apices of the lobes, 8.5 - 10.5 cm, tube 6 - 7.5 cm, very sparsely lepidote or nearly elepidote outside. Stamens 10, filaments pubescent at base. Ovary lepidote, tapered into the style; style lepidote at base. Capsule cylindric-fusiform, 4 - 5 x 1.5 - 2 cm, lepidote, strongly 5-ridged.

Notes: This species is closely related to *R. lindleyi* and *R. taggianum* but is distinguished by the bristly new growth and pubescent pedicel and usually more yellowish flowers. It is named in honour of Lady Dalhousie, wife of the former Governor General of India under the earstwhile British rule. Pradhan & Lachungpa describe a new subspecies *tashii* under

R. dalhousiae based on the colour of the flowers in some populations that grow in Sikkim; other characters are common to both.

Key to the subspecies:

1. Flowers pale creamy-white flushed with pink outside................ subsp. *dalhousiae*

1a. Flowers pale yellowish-green fading to pale yellow outside.............. subsp. *tashii*

R. dalhousiae Hook.f. subsp. *dalhousiae* exhibits two distinct varieties as under:

1. Corolla with 5 conspicuous longitudinal red lines outside................var. *rhabdotum*

1a. Corolla uniformly coloured without any redlines........................ var. *dalhousiae*

R. dalhousiae Hook.f. subsp. **dalhousiae** var. **dalhousiae. (Photo No. 33).**

Corolla usually greenish-yellow in bud fading to cream when open, or rarely white, without any longitudinal lines, slightly scented (sometimes strongly scented).

Distribution and habitat: India: Darjeeling, Arunachal Pradesh. E. Nepal, Bhutan, China (S. Xizang), 1800 - 2700 m; usually epiphytic on tree stumps or amidst rocks. **Map 7.**

R.dalhousiae Hook.f. subsp. **dalhousiae** var. **rhabdotum** (Balf.f. & R.E. Cooper) Cullen, Notes R.B.G. Edinb. 36:107 (1978). Fig. 3f, p.19. **(Photo Nos. 34, 35).**

Syn.: *R. rhabdotum* Balf.f. & R.E. Cooper, Notes R.B.G. Edinb. 10:141 (1917).

Corolla creamy white to pale yellow with 5 conspicuous longitudinal red stripes on the lobes outside. Flowering is much later than in var. *dalhousiae.*

Distribution and habitat: India : Arunachal Pradesh. Bhutan, China (S. Xizang); in the altitudinal range of 1500 - 2700 m; epiphytic or terrestrial in temperate broad-leaved mixed forests, on cliffs and amidst rocks. **Map 7.**

R.dalhousiae Hook.f. subsp. **tashii** U.C. Pradhan & S.T. Lachungpa, Sikkim-Himalayan Rhododendrons 40:1990.

This subspecies differs from subsp. *dalhousiae* in having pale yellowish-green flowers fading to pale yellow.

Disribution and habitat: India: Restricted to Sikkim growing at Pangthang at an altitude of 2000 m. **Map 7.**

Photo No. 17. *R. anthopogon* ssp.*anthopogon*.

Photo No. 18. *R. anthopogon* ssp.*anthopogon*.

Photo No. 19. *R. camelliiflorum.*

Photo No. 20. *R. cinnbarinum* ssp.*cinnabarinum.*

Photo No. 21. Deep red form of *R.cinnabarinum* ssp.*cinnabarinum* (red flowers).

Photo No. 22. *R.cinnabarinum* ssp.*xanthocodon* (orange flowers).

Photo No. 23. *R.keysii.*

Photo No. 24. *R.keysii.*

Photo No. 25. *R.edgeworthii.*

Photo No. 26. *R.pendulum.*

Photo No. 27. *R. pendulum.*

Photo No. 28. *R.glaucophyllum* var. *glaucophyllum.*

Photo No. 29. *R.glaucophyllum* var. *glaucophyllum*.

Photo No. 30. *R.maddenii* ssp. *maddenii*.

Photo No. 31. *R.maddenii* ssp.*maddenii*.

Photo No. 32. *R.maddenii* ssp.*maddenii*.

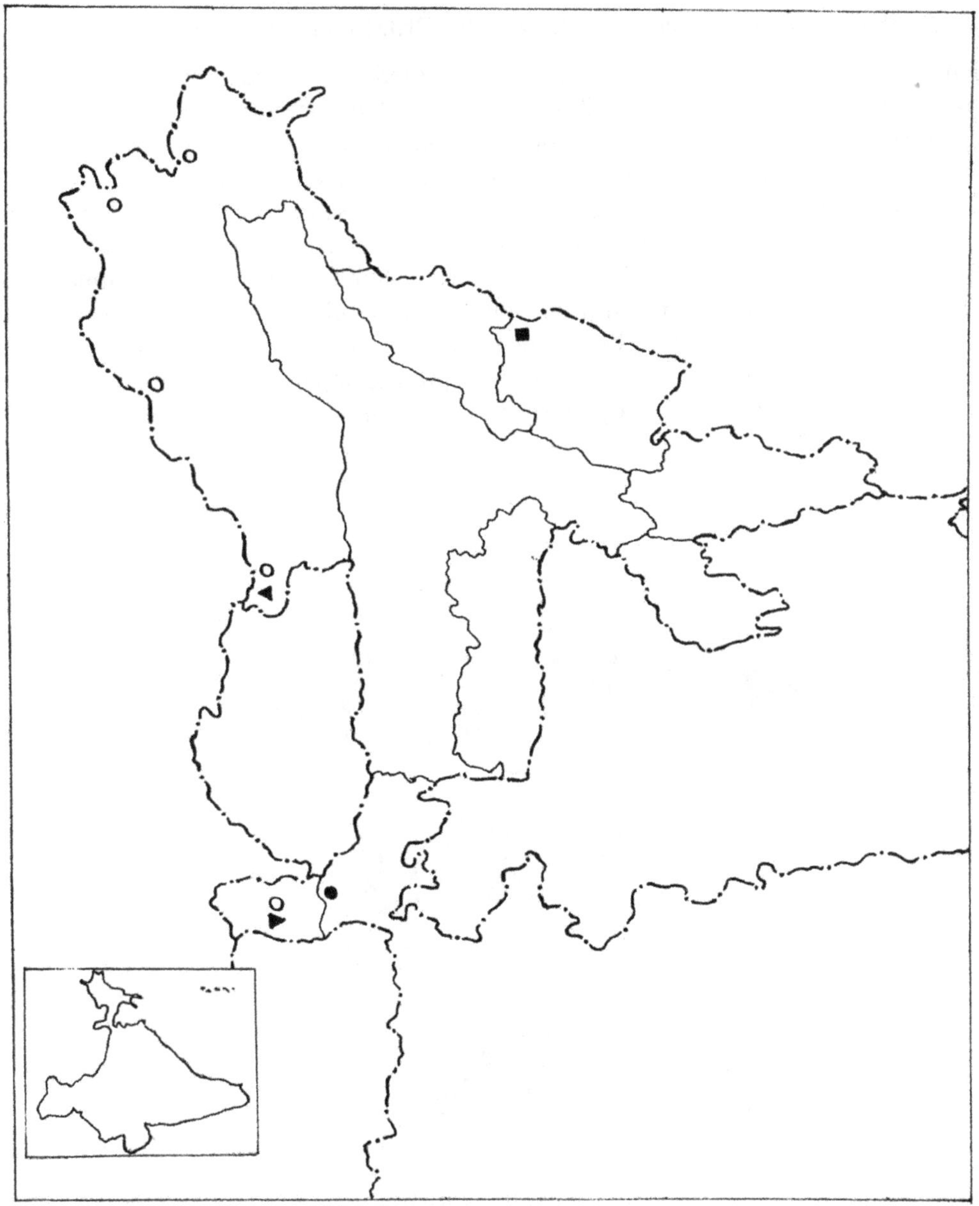

Map 7. ○ R.maddenii ssp.maddenii. ■ ssp.crassum. ● R.dalhousiae ssp.dalhousiae var.dalhousiae. ▲ var.rhabdotum. ▼ R.dalhousiae ssp.tashii.

(Maps are only indicative and not to scale)

R. lindleyi T. Moore, Gard. Chron. 1864:364 (1864). Cullen, Notes R.B.G. Edinb. 39(1): 37(1980). Cox & Cox, The Encyclopedia of Rhododendron Species 292(1997). **(Photo No. 36).**

Syn.: *R. bhotanicum* C.B. Clarke in Hook.f., Fl. Brit. India 3:475 (1882).

Epiphytic shrub, up to 4.5 m. Young shoots sometimes tinged brown or red. Bark reddish-brown, peeling. Leaves 5.5 - 12 (- 14) x 1.5 - 5.5 cm, narrowly elliptic to oblong-elliptic, rarely obovate, rounded at base, obtuse or rounded at apex, lower surface greyish-green with distant, reddish-brown, unequal scales. Inflorescence usually 3 - 7-flowered, rarely up to 12-flowered, loose; pedicels 10 - 17(25) mm, sparsely lepidote. Calyx large, deeply 5-lobed, pinkish or green, tube lepidote, the lobes narrowly obovate-oblong, 11 - 18 x 5 - 8 (10) mm, veined, hairy but elepidote. Corolla openly funnel-campanulate, 7 - 11.5 cm long, white or cream with an orange-yellow blotch at the base, sometimes tinged with pink, nearly elepidote or finely pubescent at the base, delicately fragrant. Stamens 10, pubescent towards base. Ovary lepidote, tapering into the style; style lepidote at the base. Capsule large, cylindric-fusiform, *ca* 4 cm, 5-lobed, longer than the calyx.

R.dalhousiae Hook.f.

Distribution and habitat: India : Sikkim, W. Bengal, Arunachal Pradesh, Manipur. Nepal, Bhutan, China (S. Xizang). Mostly epiphytic on large tree trunks in forests, rarely on rocks and cliffs in the altitude of 1800 - 3,400 m. **Map 8.**

Notes: *R. lindleyi* can be easily distinguished from *R. dalhousiae* in its non-bristly new shoots, more rounded flower buds, white flowers and in having hairy calyx. It is very closely akin to *R. taggianum*; only the presence of scales and absence of hairs on the calyx in *R. taggianum* separate it form *R. lindleyi.*

The species is named after Dr. John Lindley, a noted botanist and former Secretary of the Royal Horticultural Soceity, London.

R. nuttallii Booth, Kew Journ. 5:355 (1853). Cullen, Notes R.B.G. Edinb. 39(1): 36(1980). Cox & Cox, The Encyclopedia of Rhododendron Species 293(1997).

Syn.: *R. sinonuttallii* Balf. f. & Forrest, Notes R.B.G. Edinb. 13:60 (1920).

Shrub or small tree, 2 - 10 m, sometimes epiphytic. New growth sometimes reddish, brownish or purple. Leaves large, 13 - 26 x 5.5 - 13 cm, oblong–elliptic, or oblong-obovate, nearly rounded at both ends, upper surface heavily bullate, elepidote, lower surface densely covered with unequal scales, with veins very conspicuously reticulate. Inflorescence 2 - 5-flowered; pedicels 2 - 3.3 cm, lepidote. Calyx conspicuous, about 2.6 cm long, deeply 5-lobed, lobes narrowly oblong, obtuse, sometimes with a few filiform-acicular hairs. Corolla funnel-campanulate, margins sometimes wavy or reflexed, very large, up to 12.5 cm long and across, white to pink-tinged with a yellow blotch towards the base, fragrant, mouth oblique, tube 7 - 8 cm, sparsely lepidote outside. Stamens 10, filaments pubescent at base. Ovary densely lepidote, tapered into the style. Capsule densely lepidote, ovoid cylindric, 5 - 7 x 1.8 - 2 cm, faintly 5-ridged, much longer than the persistent calyx.

Distribution and habitat: India : Arunachal Pradesh. Bhutan, China (N.W. Yunnan, S.E. Xizang). **Map 8.**

Notes: This species grows in the altitudinal range of 1,100 - 3,700 m, often as an epiphyte or in thickets, forest margins adjoining swampy habitats, rocky slopes and cliffs. The senior author (A.R.K.S.) collected this species in Subansiri district, Arunachal Pradesh as an epiphyte along forest margin adjacent to an open *Sphagnum* bog in S.E. direction of Hapoli, in April 1965. Rather rare.

The species is easily identified by its large wrinkled leaves with deeply impressed venation on the upper surface and by its very large fragrant, beautiful flowers, which are one of the largest in the genus *Rhododendron.*

It is named after Thomas Nuttall, a botanist and traveller.

R. taggianum Hutchinson, Notes R.B.G. Edinb. 16:178 (1931). Cullen, Notes R.B.G. Edinb. 38(1) : 38(1980). Cox & Cox, The Encyclopedia of Rhododendron Species 294(1997).

Syn.: *R. headfortianum* Hutchinson, Bot. Mag. 163: t. 9614 (1942).

Shrub, 2-3 m high. Bark dark reddish-brown and grey, peeling. Leaves up to 12.5 x 6 cm, oblong-lanceolate to ovate, lower surface with small, glandular, unequal, distantly arranged scales. Inflorescence (1 -) 3 - 5-flowered; pedicels up to 2 cm, sparsely lepidote. Calyx large, *ca* 2 cm, deeply 5-lobed, not hairy, often with deciduous scales on the margins, corolla openly tubular funnel-shaped, 5.5 - 9.5 cm, cream coloured in bud, opening pure white, throat yellow-blotched, highly fragrant. Stamens 10. Ovary lepidote, tapering into the style. Capsule 5-ridged, *ca* 4 cm, much longer than the calyx lobes.

Distribution and habitat: India : Arunachal Pradesh. N.E. Burma, China (N.W. Yunnan, S.E. Xizang). In conifer forest margins, rocky slopes, thickets or sometimes epiphytic on evergreen trees in the altitude of 1800 - 3700 m. **Map 8.**

Notes: *R. taggianum* is very similar to *R. lindleyi* and perhaps should be treated as a variety of it but differs from *R. lindleyi* in its seldom epiphytic habit, and in its non-hairy but often lepidote margins of the calyx lobes, typically more elongated flower buds with wider corolla and broader leaves. It is also stated to be less hardier with better root system and more spectacular with very fragrant flowers than *R. lindleyi.*

The species is named after H.F. Tagg, a former botanist in the Royal Botanical Garden, Edinburgh.

Group 3: Megacalyx

This group is characterized by deeply pitted leaf scales; petioles with a shallow groove; large, deeply 5-lobed calyx; stamens 10; ovary 5-locular; pedicels divaricate-reflexed in fruit.

R. megacalyx Balf. f. & Kingdon Ward, Notes R.B.G. Edinb. 9:246 (1916). Cullen, Notes R.B.G. Edinb. 39(1): 41(1980). Cox & Cox, The Encyclopedia of Rhododendron Species 295(1997).

Shrub or small open tree with loose branches, 2 - 7.5 m high, rarely epiphytic. Bark light brown, peeling, smooth. Leaves up to 19 cm long, 7.5 cm broad, elliptic to obovate-elliptic, rounded at apex, upper surface bullate, lower surface with prominently sunken veins, glaucous and densely lepidote with heteromorphic, sub-contiguous, golden or brownish scales, the smaller ones deeply sunk in pits; petiole shallowly grooved on the upper surface. Inflorescence 2 - 6-flowered; pedicels up to 3.5 cm long, accrescent and divaricate-reflexed in fruit, more or less elepidote. Calyx very large, lobed to the middle, cup-like, green or pinkish-green, lobes broadly elliptic, rounded at apex, 2.2 - 3 x 1.4 - 2 cm, elepidote. Corolla funnel-campanulate, mouth very oblique, lower lobes larger, 7.5 - 11 cm long, often flushed purple in bud, opening pure white or cream coloured, with a nutmeg-like scent, tube 5 cm long, very sparsely lepidote. Stamens 10, filaments pubescent towards base. Ovary densely lepidote, tapering into the style. Capsule 2 - 2.5 x 1 - 1.5 cm, cylindric-globose, shorter or as long as the calyx.

Distribution and habitat: India : Arunachal Pradesh. N.E.Myanmar, China. In the altitudes of 1800 - 4000 m, usually in forests and thickets, on steep cliffs and along water courses. **Map 8.**

Notes: This is a distinctive species easily identified by its sunken leaf veins and pitted smaller golden or brownish scales, channelled petiole, large elepidote calyx, larger lower corolla lobes and the small capsule equalling or shorter than the calyx. This species is rare and collected from Lohit valley and Mishmee Hills in the eastern most part of Arunachal Pradesh. The species derives its name from its large calyx.

Group 4. Ciliicalyx / Johnstoneanum Alliance

This alliance is characterized by leaves with main vein totally impressed above; pedicels and calyx not covered with waxy whitish bloom; pedicel not divaricate-reflexed in fruit; calyx often small, ciliate along margins; stamens (8 -) 10, and with 5-celled ovary.

This Alliance which is based on the *Ciliicalyx* subseries is considered taxonomically most problematic. Though Cullen and Sleumer reduced the number of species in this group, Cox and Cox (*The Encyclopedia of Rhododendron Species 195: 1997*) emphasize the need for more fieldwork for better understanding and a definitive reclassification. Cullen (1980) makes three subgroups within this Alliance, namely, a large group which includes the yellow-flowered species and *R. formosum* and its relatives; the second group centered around *R. johnstoneanum*; and the third to deal with *R. ciliicalyx*. The treatment outlined by J. Cullen (*Notes R.B.G. Edinb. 39(1):41 – 50 (1980)*) and followed by Cox and Cox in their above reference is considered in the present work.

4A. Shrubs, under 2 m tall; style impressed.

R. ciliatum Hook.f., Rhododendrons of Sikkim Himalaya t. 24 (1849). Cullen, Notes R.B.G. Edinb. 39(1): 41(1980). Cox & Cox, The Encyclopedia of Rhododendron Species 296(1997). **(Photo No. 37).**

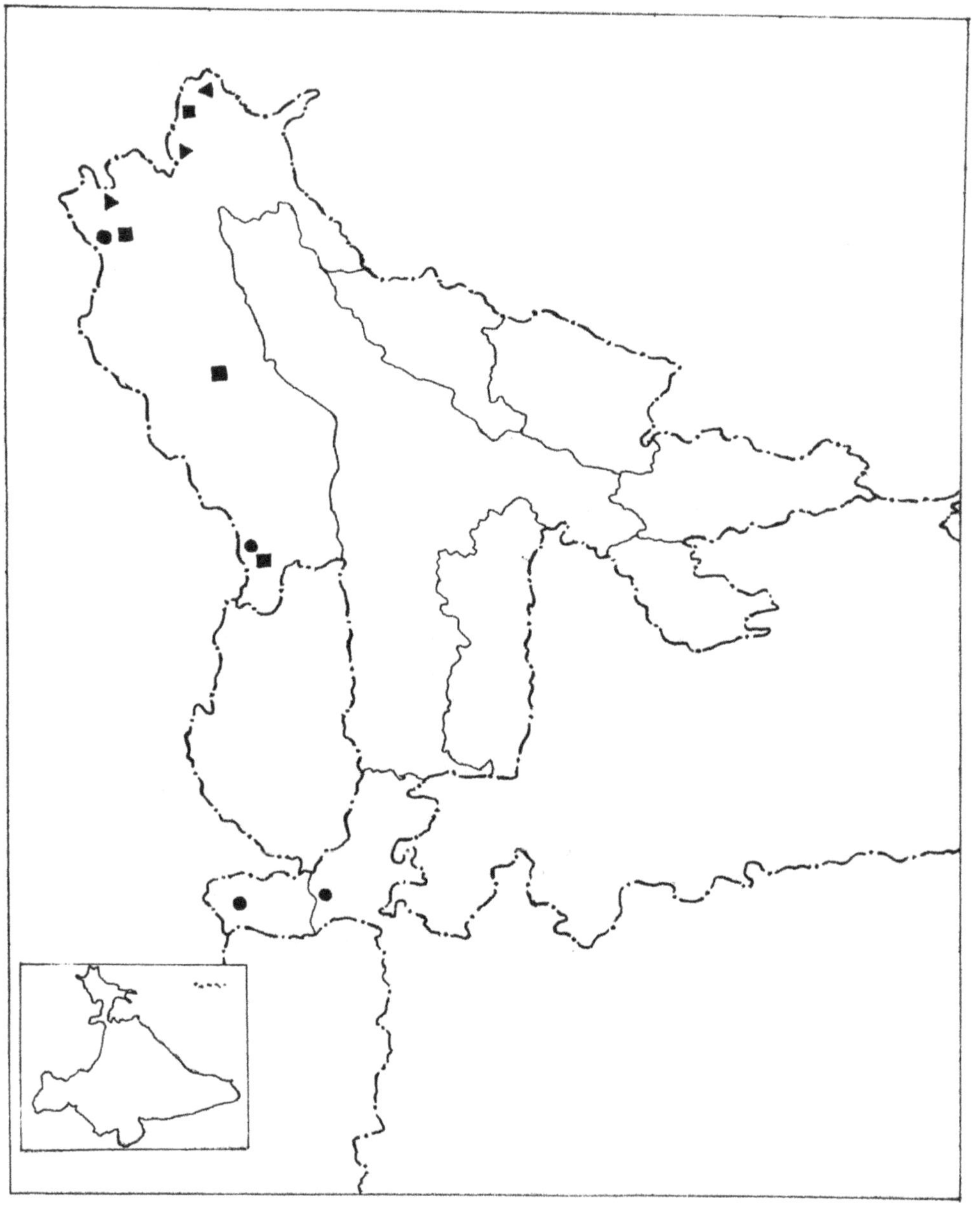

Map 8. ● R.lindleyi. ■ R.nuttallii. ▲ R.taggianum. ▼ R.megacalyx.

(Maps are only indicative and not to scale)

Shrub, 2 m high. Branchlets hairy. Bark rough, reddish-brown, peeling. Leaves 4 - 9 x 2 - 3.5 cm, elliptic to narrowly-elliptic, margins hairy, upper surface sparsely loriform-setose, lower surface scattered with brown, unequal scales. Inflorescence 2 - 5-flowered; pedicels lepidote, densely loriform-setose. Calyx conspicuous, usually 5-lobed, lobes oblong-ovate, 6 - 9 mm, ciliate on the margins. Corolla campanulate or funnel-campanulate, 2.5 - 5 cm long, white to pale pink, glabrous. Stamens 10, filaments pubescent at base. Ovary lepidote; style impressed, glabrous and elepidote. Capsules oblong-globose, 1 - 1.6 cm, lepidote.

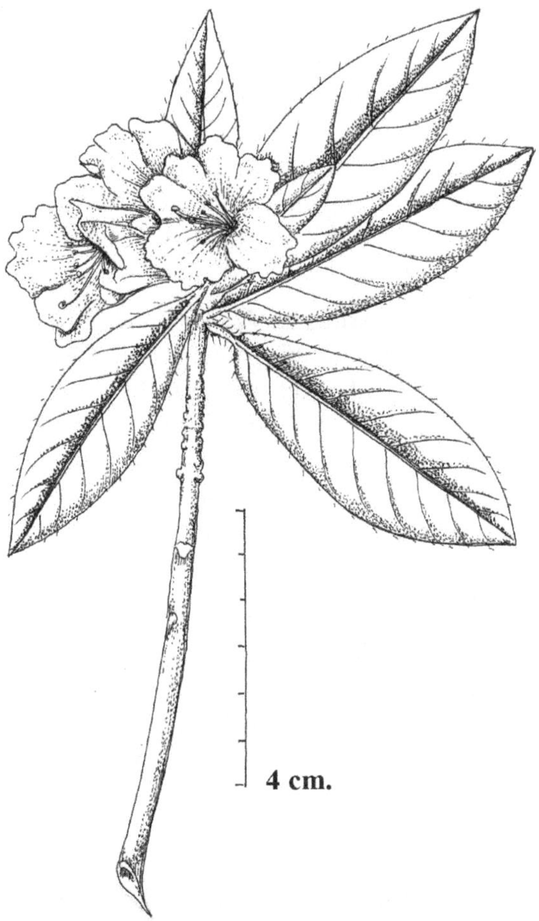

R. cilliatum Hook.f.

Distribution and habitat: India : Sikkim. E. Nepal, Bhutan, China (S. Xizang). Frequent in conifer-rhododendron forests along hill sides, rocky slopes and water courses in the altitudinal range of 2,400 - 4,000 m. **Map 9.**

Notes: This species is easily identified in the field by its hairy leaves particularly along the leaf margins, petiole and calyx and by its white to pale pink flowers. The Himalayan stock of this species introduced into U.K has been a successful parent of hybrids such as '*R. 'Praecox'* and *'R. Cilpinense'*. Its very variable flower quality and hardiness have been taken much advantage of in introducing this species outside its distributional localities and wild habitats. The species name *ciliatum* refers to its leaf-margins fringed with hairs.

R. coxianum Davidian, Rhododendrons (R.H.S.) 51 (1972). Cullen, Notes R.B.G. Edinb. 39(1): 57(1980). Cox & Cox, The Encyclopedia of Rhododendron Species 297(1997).

Epiphytic or straggling shrub, to 3 m high. Branches bristly hairy. Leaves 5 - 11 x 1.5 - 3 cm, oblanceolate, nearly rounded at base, lower surface with unequal, transluscent to brown scales, scales apart about 2.5 times their own diameter; petiole with or without wings at margins. Inflorescence 2 - 4-flowered; pedicels lepidote. Calyx unequal, more or less hairy. Corolla tubular-funnel-campanulate, up to 7 cm long, creamy-white, greenish-yellow blotched at base, lepidote outside, strongly scented. Stamens 10. Ovary 7-celled, densely lepidote; style impressed, lepidote in the lower part, long and slender.

Distribution and habitat: India : Arunchal Pradesh, *ca.* 15 km S.E. of Apa Tani valley, Subansiri district. In forest margins adjoining an open marshy / boggy area and epiphytic on forest tree trunks at an altitude of 1700 m (5,500 ft). The species was first collected from the above locality in April 1965 by P.A. Cox & P. Hutchinson and the senior author(ARKS) during a joint botanical exploration of the Botanical Survey of India and the specimens were only in very young flower buds and found to be very rare in the wild. Live plants of the species introduced into Glendoick gardens of the Coxes at Glencarse near Perth in Scotland flowered in 1971 based on which the species was described as new naming it after Mr. P.A. Cox. **Map 9.**

Notes: J. Cullen (Notes R.B.G. Edinb. 39(1):57 (1980) treats this as an imperfectly known species of Ciliicalyx aggregate in the subsection *Maddenia* and considers it to be a variant of *R. formosum.* While treating it to be a separate species, Cox & Cox (*The Encyclopedia of Rhododendron Species* 297, 1997) express that *R. coxianum* may be an extreme form of *R. formosum* closest to var. *inaequale* in having flowers with a similar scent, but differing in the stiff leathery leaves and the non-frilled corolla lobes and in having the lower two corolla lobes that are sometimes split. Living specimens in flower of the two taxa in question make one to be inclined to treat *R. coxianum* as a sepearate species rather than a mere variant of *R. formosum* var. *inaequale*,though they look close in herbarium specimens.

R. formosum Wallich, Plant. Asiat. Rar. 3(3):t.207 (1832). Cullen, Notes R.B.G. Edinb.39(1): 46(1980). Cox & Cox, The Encyclopedia of Rhododendron Species 298(1997).

Straggly shrub, 3 m high. Young branchlets bristly. Leaves 3 - 12 x 1 - 5 cm elliptic or oblanceolate-obovate, margins fringed with white hairs, tapering at base, acute or acuminate at apex, upper surface dark green, elepidote, lower surface with unequal scales about their own diameter apart; margins loriform setose; petioles bristly. Inflorescence 2 - 6-flowered ; pedicels lepidote. Calyx disciform, scaly and sparsely hairy. Corolla widely funnel-shaped, 4.5 - 7.5 cm long, white to white tinged with pink, often with a yellow blotch and pink margins or stripes along the tube outside, tube pilose at the base inside and lepidote outside, scented. Stamens 10, filaments pubescent near the base. Ovary lepidote, style impressed, lepidote towards the base. Capsule *ca.* 1.6 cm, lepidote.

Distribution and habitat: India : North East India : Meghalaya, Manipur, Arunachal Pradesh. Frequent along forest margins, rocky situations, and river banks in the altitudes of 600 - 1800 m.

Notes: This species is very variable with a wider distribution in India and has been segregated into two varieties:

1. Leaves narrow, to 1.5 cm broad; corolla often flushed pink, delicately scented .... var. *formosum*

1a. Leaves 1.5 - 5 cm broad ; corolla white with yellow blotch, very fragrant than var. *formosum* .... var. *inaequale*

R. formosum Wallich var. **formosum**

Syn.: *R. gibsoni* Paxton, Mag. of Bot. 8: t.217 (1841). *R. formosum var. salicifolium* C.B.Clarke in Hook.f., Fl. Brit. India 3:473 (1882). *R. iteaphyllum* Hutchinson, Notes R.B.G. Edinb. 12:83 (1919).

Distribution and habitat: India : N.E. India : Meghalaya. Endemic. In pine mixed evergreen sub-tropical forests in the altitudes of 1000 - 1500 m. Flowering during April-June. **Map 9.**

R. formosum var. **inaequale** (Hutchinson) Cullen, Notes R.B.G. Edinb. 36:108 (1978).

Syn. : *R. inaequale* Hutchinson, Notes R.B.G. Edinb. 12:75 (1919).

Distribution and habitat: India : Meghalaya, Manipur, Arunachal Pradesh. Endemic. In mixed sub-tropical evergreen forests in the altitudes of 1000 – 1500 m. Flowering during March - May. **Map 9.**

Notes: The var. *formosum* is hardier and grows well even in open conditions amongst rocks while var. *inaequale* apparently is less hardier and tender and prefers somewhat shady situations in forested hill slopes and along forest margins to grow. Its flowers are very sweet smelling making it one of the finest shrub in the subsection *Maddenia.*

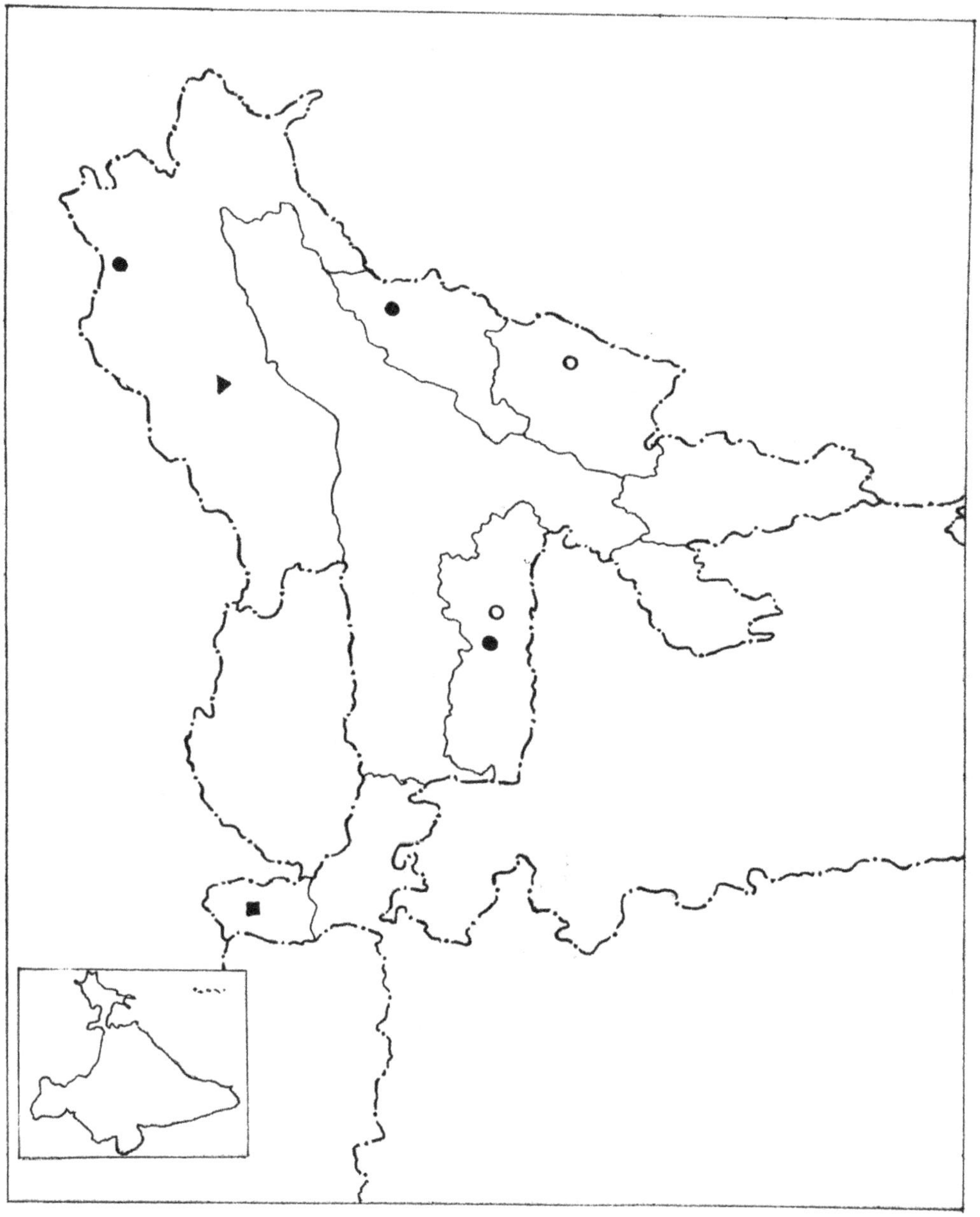

Map 9. ■ R.ciliatum. ▼ R.coxianum. ● R.formosum var.formosum. O var.inaequale.

(Maps are only indicative and not to scale)

4B. Johnstoneanum Aggregate

Large shrubs. Corolla large; style impressed.

R. dendricola Hutchinson, Notes R.B.G. Edinb. 12:60 (1919). Cullen, Notes R.B.G. Edinb. 39(1):48(1980). Cox & Cox, The Encyclopedia of Rhododendron Species 301(1997).

Syn.: *R. atentsiense* Handel-Mazzetti, Anz. Akad. Wiss. Wein 18:14 (1921). *R. notatum* Hutchinson, Notes R.B.G. Edinb. 16:177 (1931).

Shrub, terrestrial or epiphytic, up to 4.5 m high. Young branches and growth rarely loriform-setose, bark smooth, dark purple to mahogany-red, peeling. Leaves 7 - 12 x 3 - 5 cm, narrowly elliptic to narrowly obovate, narrowed to the base, abruptly acute or shortly acuminate at apex, lower surface scaly of variable density. Inflorescence 2 - 5-flowered. Calyx disc-like or obscurely lobed, not hairy. Corolla funnel-shaped, up to 10 cm long, white often with a greenish, yellow or orange blotch towards the base, or flushed pink, usually fragrant, lepidote outside, tube usually pilose at base. Stamens 10. Ovary lepidote; style impressed. Capsule up to 2 cm long.

Distribution and habitat: India : Arunachal Pradesh. N.E. Burma, China (N.W., W. & C. Yunnan, S.E. Xizang). Rare on rocks, cliffs, forest margins and slopes. **Map 10.**

Notes: "This is a variable species, differing from *R. johnstoneanum* in the non-bristly foliage, and from *R. walongense* in the absence of calyx hairs". (Cox & Cox, *The Encyclopedia of Rhododendron Species* 302: 1997).

The species name *dendricola* means "dweller on trees" indicating its often epiphytic habit.

R. johnstoneanum (Watt ex) Hutchinson, Notes R.B.G. Edinb. 12:72 (1919). Cullen, Notes R.B.G. Edinb. 39(1): 47(1980). Cox & Cox, The Encyclopedia of Rhododendron Species 302(1997).

Syn.: *R. formosum* var. *johnstoneanum* Watt ex Brandis, Indian Trees 411 (1906). *R. parryae* Hutchinson, Gard Chron. 93:386 (1933).

Shrub, up to 4.6 m, growing broader than height with age. Young branches loriform setose, aromatic. Leaves 5 - 10 x 2.4 - 3 cm, elliptic to broadly elliptic, tapering to the base, obtuse or subacute at apex, bristly along the margins, lower surface densely scaly, scales contiguous or overlapping, brown; petiole hairy and scaly. Inflorescence 2 - 5-flowered. Calyx disc-like, long hairy on the margins. Corolla 4.5 - 6 cm long, funnel-campanulate, white, often with a yellowish blotch and pink or purplish tinge, sometimes fragrant, corolla tube 2.5 - 3 cm long, pilose at the base, laxly lepidote all over; pedicel densely lepidote. Stamens 10, filaments pubescent towards the base. Ovary lepidote; style impressed, lepidote. Capsule 1.6 - 2.2 cm long, lepidote.

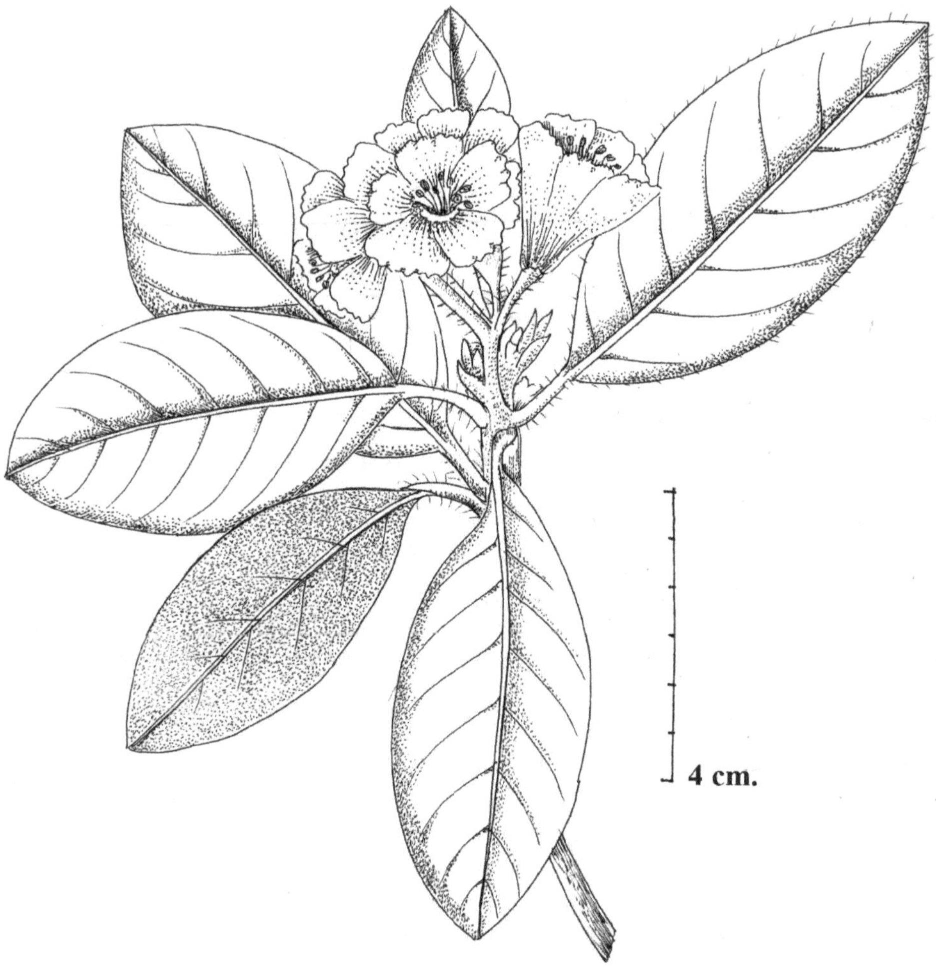

R. johnstoneanum Watt ex Hutchinson

Distribution and habitat: India : Manipur and Mizoram. Endemic. Frequent, in open scrub or forest margins, mountain slopes, sometimes epiphytic; in the altitudinal range of 1850-3100 m (6000-10000 ft). This is a variable species with bristly / hairy branchlets, petiole and young leaves, lower leaf surface with dense, contiguous or overlapping brownish scales and usually creamy-yellow flowers. **Map 10.**

R. veitchianum Hook., Bot. Mag. 83: t.4992(1857). Cullen, Notes R.B.G. Edinb. 39(1): 55(1980). Cox & Cox, The Encyclopedia of Rhododendron Species 307(1997).

Epiphytic or terrestrial shrub up to 3.5 m. Bark brown, flaking. New growth sparsely loriform-setose. Leaves 6 - 10 x 2 - 4 cm, obovate or narrowly elliptic, tapering at both ends and acuminate at apex, upper surface dark green, lower surface pale green with distinct, unequal, brown scales; petiole lepidote. Inflorescence 2 - 5-flowered; pedicels lepidote. Calyx disc-like, lepidote, loriform-ciliate. Corolla openly funnel-campanulate, 5 - 6.5 cm long, about 12 cm wide, white, often with a yellow-blotch; tube 2.5 - 4 cm long; lobe margins frilled or wavy. Stamens 10, pubescent at base. Ovary lepidote, tapering into the style. Capsule lepidote, *ca* 3 cm long.

Distribution and habitat: India : Mizoram in N.E. India. Burma, Thailand and Laos. Rare, usually epiphytic in forests, between 1200 - 1700 m altitude. **Map 10.**

R. walongense Kingdon-Ward, Gard. Chron. 133:5 (1953). Cullen, Notes R.B.G. Edinb. 39(1):48(1980). Cox & Cox, The Encyclopedia of Rhododendron Species 303(1997).

Shrub, 2 - 3 m, terrestrial or sometimes epiphytic. Young branches not hairy. Leaves 10 - 11 x 3.8 - 4.5 cm, elliptic, tapering to the base, acute at apex, lower surface brownish, with lax, unequal scales. Inflorescence 3 - 4-flowered. Calyx small, disc-like, undulately lobed, lepidote and hairy along the margins. Corolla funnel-shaped, 6 - 7 cm long, creamy-white with a greenish blotch, densely pilose all over, the tube *ca* 4 cm long, the lobes sparsely lepidote. Capsule *ca* 1.5 cm long.

Distribution and habitat: India : Arunachal Pradesh. China (S.E. Xizang). Rare, in mixed forests amidst rocks, cliffs and ravines, rarely epiphytic, in the altitudinal range of 1500 - 2100 m. **Map 10.**

Notes: This species is distinguished by its dense pubescence all-over the corolla outside and persistent hairs on the margins of calyx lobes. A very interesting form was collected by Peter A. Cox & P.Hutchinson and by one of the authors (ARKS) from Subansiri district, Arunachal Pradesh during their joint botanical exploration in April, 1965. This was previously identified as *R. parryae.* **(Photo No. 38).** It was epiphytic in the wild, with deep mahogany-coloured, smooth, peeling bark and pale creamy-white flowers with a characteristic spicy scent. This form is in cultivation in the Glendoick gardens at Glencarse, Perth Shire, Scotland.

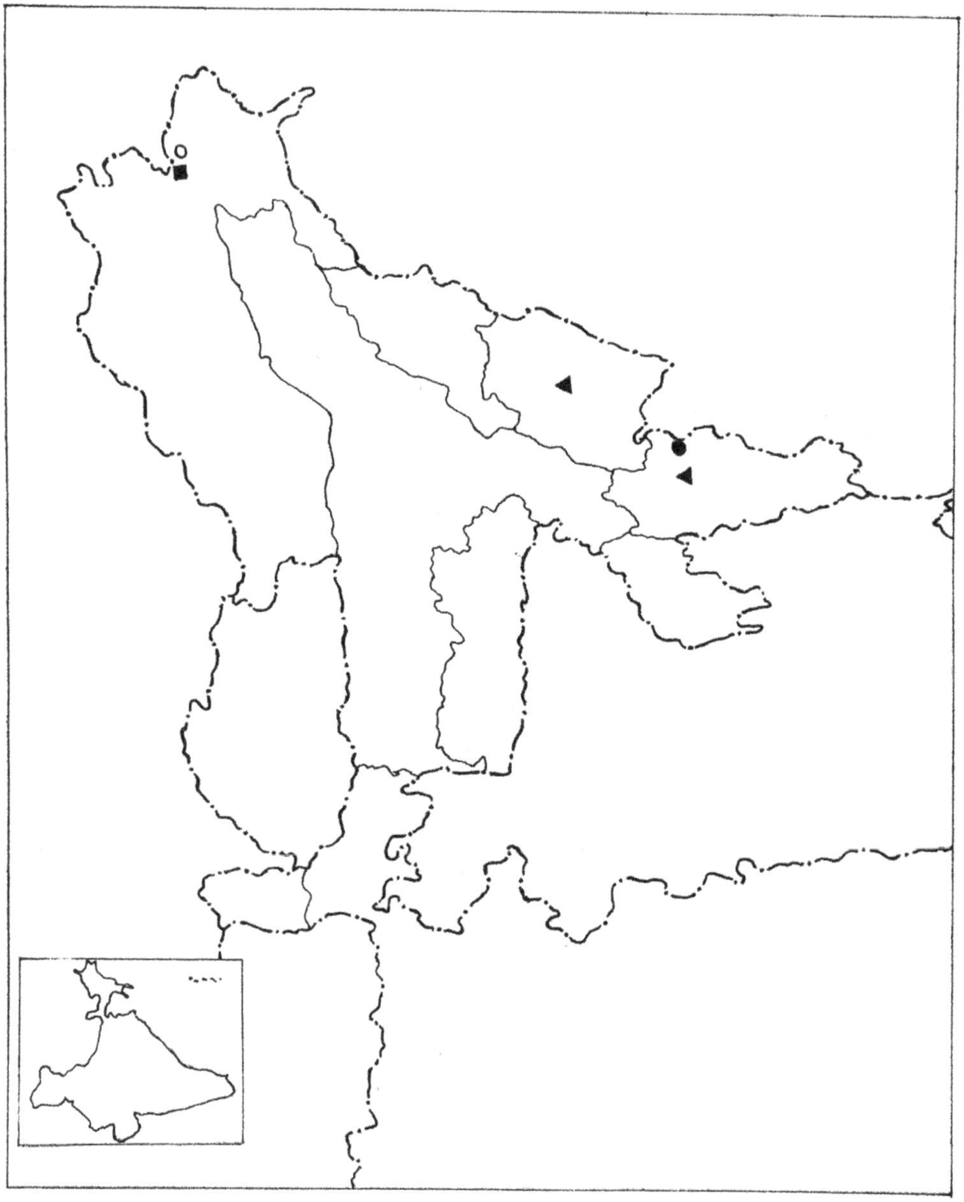

Map 10. ■ R.dendricola. ▲ R.johnstoneanum. ● R.veitchianum. O R.walongense.

(Maps are only indicative and not to scale)

Subsection Monantha

Subsection **Monantha** Cullen, Notes R.B.G. Edinb. 36:122 (1978).

Epiphytic or terrestrial shrubs. Leaves evergreen. Scales large, flat, often unequal, broadly rimmed. Inflorescence terminal, 1 - 3-flowered. Calyx obscure, or 5-lobed. Corolla lepidote, yellow or purple, tubular-funnel shaped to tubular-campanulate, lobes scarcely spreading. Stamens 10, filaments pubescent towards base. Stamens and style exerted beyond corolla lobes. Ovary 5-locular; style impressed. Capsule lepidote. Seeds winged and with finned margins.

Type species: *R. monanthum* Balf. f. & W.W. Smith

This subsection is related to subsections *Boothia* and *Uniflora*. The distinctive corolla shape and exerted stamens and style and the winged and finned seeds are identifying features which are common to section *Vireya*, and is apparently a link between sections *Vireya* and *Rhododendron*.

Key to the species of subsection Monantha

1. Inflorescence 1(- 2)-flowered; corolla pinkish-purple or purple........ *concinnoides*

1a. Inflorescence 3-flowered; corolla bright yellow .... *kasoense*

R. concinnoides Hutchinson & Kingdon-Ward, Notes R.B.G. Edinb. 16:180 (1931). Cullen, Notes R.B.G. Edinb. 39(1): 61(1980).

Epiphytic shrub up to 3 m high. Young growth densely lepidote. Leaves 2.5 - 5 cm long, elliptic or ovate-elliptic, densely covered with dark brown scales on the lower surface. Inflorescence 3-flowered. Corolla *ca* 2.5 cm long, tubular-campanulate, pinkish-purple or purple-coloured, with darker spotting at base, scaly outside. Stamens 10. Ovary 5-celled; style impressed.

Distribution and habitat: India : Arunachal Pradesh. Endemic. Often epiphytic on conifer trees in rhododendron mixed forests in the altitudes of 2700 - 3350 m. Very rare. **Map 11.**

Notes: This species is perhaps known only from the Type collection from Delei Valley, at 2700 - 3750 m alt., essentially differing from *R. monanthum* only in its purple coloured flowers.

R. kasoense Hutchinson & Kingdon-Ward, Notes R.B.G. Edinb. 16:181 (1931).

Epiphytic shrub up to 1 m high. Young growth densely lepidote. Leaves 3 - 4.5 x 1.4 - 2 cm, ovate-elliptic or elliptic, cuneate at base, acute at apex, upper surface dark green, lower surface pale and sparsely papillose hairy with distantly placed scales. Inflorescence 3-flowered; pedicels densely scaly. Calyx rim-like, densely scaly. Corolla tubular-funnel shaped or tubular-campanulate, 1.4 - 2 cm long, bright yellow, scaly outside. Stamens 10. Ovary lepidote; style impressed, longer than stamens.

Distribution and habitat: India: Arunachal Pradesh. China (S.E. Xizang). In forests, in the altitudes between 2450 - 2750 m. **Map 11.**

Notes: This species is known only from a few collections made by Kingdon-Ward from Kasop (2300 - 2700 m) in Delei Valley, close to Indo-Tibetan border. It is very similar to *R. monanthum* differing in its less clearly papillose lower leaf surface with paler, distantly placed scales and more importantly in its 3-flowered inflorescence with bright yellow corolla.

Subsection Saluenensia

Subsection **Saluenensia** (Hutchinson) Sleumer, Bot. Jahrb. 74:534 (1949). Cullen, Notes R.B.G. Edinb. 39(1): 114(1980). Cox & Cox, The Encyclopedia of Rhododendron Species 316(1997).

Syn.: Series *Saluenense* sensu Hutchinson in Stevenson (ed.) The Species of Rhododendron 587 (1930) & sensu Davidian in R.H.S. Rhodo. Yearbook 8: 84 – 98 (1954).

Prostrate, bushy to erect shrubs, reaching 1.5 m high. Young growth lepidote and loriform-setose, the setae persistent. Leaves evergreen, lower surface densely lepidote with overlapping, crenulate scales. Inflorescence 1 - 3 (rarely 7)-flowered, terminal. Calyx large, deeply 5-lobed, lobes usually loriform-ciliate. Corolla openly-funnel-campanulate, pink, reddish-crimson, magenta, purple, purplish-crimson, pilose and lepidote outside, pilose within the tube. Stamens 10, declinate, filaments densely villous-pilose in the lower half. Ovary lepidote, 5-locular; style impressed, declinate, reddish in colour, glabrous and without scales. Capsule lepidote, small, shrivelled. Seeds without wings, obscurely finned.

Type species: *R. saluenense* Franchet

This subsection is represented in India by *R. calostrotum* Balf.f. & Kingdon-Ward subsp. *riparium* (Kingdon-Ward) Cullen.

Cullen (Notes R.B.G. Edinb. 39(1): 114 (1980) reduced the eight species that were earlier included in this subsection to two, namely, *R. calostrotum* and *R. saluenense* which is followed in this book. Cox & Cox (*The Encyclopedia of Rhododendron Species* 316, 1997) opine that the two species could even be merged into a single one as there are integrades between *R. saluenense* and *R. calostrotum.*

Species in the subsection are characterized by the widely funnel-campanulate corolla and the overlapping scales underneath the leaves.

R. calostrotum Balf.f. & Kingdon-Ward subsp. **riparium** (Kingdon-Ward) Cullen, Notes R.B.G. Edinb. 36:112 (1978).

Syn.: *R. rivulare* Kingdon-Ward, Gard. Chron. 86:503 (1929). *R. riparium* Kingdon-Ward, Notes R.B.G. Edinb. 16:180 (1931). *R. calciphilum* Hutchinson & Kingdon-Ward, Notes R.B.G. Edinb. 16:179 (1931).

Prostrate shrub up to 1.5 m high. Young growth densely lepidote, not hairy or bristly or the hairs soon deciduous. Leaves 1 - 3.5 x 0.4 - 2 cm, suborbicular to oblong-ovate, margins sparsely loriform-ciliate, upper surface rarely glaucous with persistent, dried-out scales, lower surface with dense, overlapping scales usually arranged in 3 - 4 distinct tiers, the outermost tier with long stalks and cup-shaped discs. Inflorescence 2 - 5-flowered ; pedicels 1 - 1.5 cm long with stalked scales. Calyx lobes oblong to obovate, scaly and fliform-acicular pubescent, margins hairy. Corolla 1.8 - 2.8 cm long, pinkish-purple. Stamens 10. Ovary lepidote, without hairs. Capsule 0.6 - 0.9 cm long, lepidote.

Distribution and habitat: India : Arunachal Pradesh. N.E. Burma, China (S. & S.E. Xizang, N.W. Yunnan). In open rocky slopes, alpine meadows and in damp, swampy areas. Rare. Kingdon-Ward's Coll. No.8229 cultivated in the Royal Botanical Garden, Edinburgh was collected by him from Mishmee hills, Lohit District, Arunachal Pradesh bordering S.E. Tibet. **Map 11.**

Notes: *calostrotum* means 'with a beautiful covering', refering to its dense covering of scales on young growth and leaves.

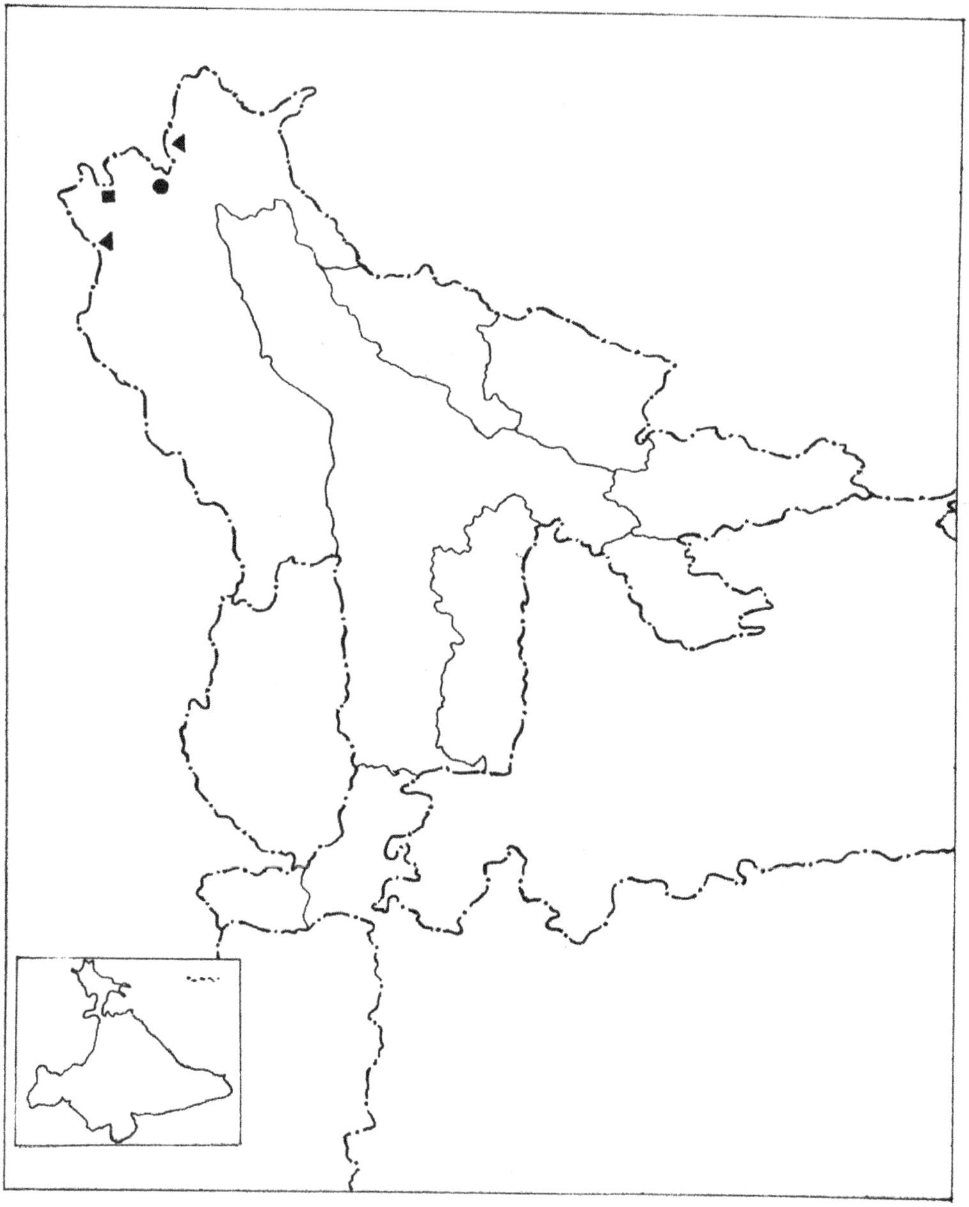

Map 11. ● R.concinnoides. ■ R.kasoense. ▲ R.calostrotum ssp. riparium.

(Maps are only indicative and not to scale)

Subsection Tephropepla

Subsection **Tephropepla** (Cowan & Davidian) Sleumer, Bot. Jahrb. 74:532 (1949).

Syn.: Series *Boothii* sensu Hutchinson in Stevenson (ed.) The Species of Rhododendron 156 (1930) pro parte.

Series *Boothii* subseries *Tephropeplum* Cowan & Davidian, Rhodo. Yearbook (RHS) 3:72 (1948) pro max. parte.

Dwarf, compact, erect or spreading shrubs, up to 2 m high. Bark sometimes peeling. Leaves evergreen, lanceolate to oblong-obovate or oblanceolate; lower surface papillose and lepidote. Inflorescence usually terminal. Calyx deeply 5-lobed, lobes erect or reflexed. Corolla campanulate to tubular-campanulate, pink, red, white, cream or yellow, sometimes pubescent outside. Stamens 10, filaments pubescent. Ovary 5-locular, lepidote; style long, slender, tapering or impressed. Capsule lepidote.

Type species: *R. tephropeplum* Balf.f. & Farrer

Notes: This subsection is related to subsection *Boothia* (via *R. tephropeplum* and *R. xanthostephanum* that were earlier included in the subseries *Tephropeplum* under Series *Boothii*).

Key to the species of subsection Tephropepla

1. Corolla white, red or pink; style impressed on to the ovary.. *tephropeplum*

1a. Corolla cream or yellow; style tapering from top of the ovary .... *xanthostephanum*

R. tephropeplum Balf.f. & Farrer, Notes R.B.G. Edinb. 13:302 (1922).

Syn.: *R. spodopeplum* Balf.f. & Farrer, op. cit.:299. *R. deleiense* Hutchinson & Kingdon-Ward, Notes R.B.G. Edinb. 16:172 (1931). Cox & Cox, The Encyclopedia of Rhododendron Species 329 (1997).

Shrub, 0.5 - 1.3 m (rarely to 1.8m). Bark brownish, peeling. Leaves 3 - 13 x 1 - 4 cm, narrowly oblanceolate to narrowly elliptic to oblanceolate, rounded at the apex, cuneate at the base, dark green on the upper surface, densely lepidote and glaucous on the lower surface, scales dark, unequal, sunken in pits. Inflorescence 3 - 9-flowered; pedicels lepidote, 1.6 - 1.8 (– 2) cm long. Calyx large, leafy, 5 - 7 mm, lepidote and setose. Corolla campanulate to tubular-campanulate, 1.7 - 3.2 cm long, differently shaded from pink to red, purplish, rarely white; tube 1.1 - 1.8 cm, sparsely lepidote outside. Stamens 10, filaments pubescent in the lower half. Ovary lepidote; style longer than the corolla, lepidote in the lower half, impressed on the ovary. Capsule ovoid-cylindric, 0.7 - 1 cm.

Distribution and habitat: India : Arunachal Pradesh (Delei valley, Mishmee Hills, Lohit District). N.E. Burma, China (S.E. Xizang, N.W. Yunnan). Rather uncommon and it seems to be restricted to eastern Arunachal Pradesh in India. Grows on rocks and cliffs and in alpine meadows in the altitudinal range of 2,400 - 4,300 m. **Map 12.**

Notes: The species derives its name *tephropeplum* meaning 'with ash-grey covering', from its glaucous undersurfaces of leaves.

R. xanthostephanum Merrill in Brittonnia 4:148 (1941). Cullen, Notes R.B.G. Edinb. 39(1): 127(1980). Cox & Cox, The Encyclopedia of Rhododendron Species 330(1997).

Syn.: *R. aureum* Franchet, Journ. de Bot. 9:394 (1895), non Gregori.

Shrub, erect or spreading, up to 3 m. Bark smooth, reddish-brown. Leaves evergreen, 5 - 10 x 1 - 3.4 cm, oblong, narrowly-elliptic to elliptic, lower surface glaucous, scales unequal, smaller deeply sunk in pits, ½ to their own diameter apart. Inflorescence 3 - 5-flowered, terminal, rarely axillary; pedicels 1 - 1.2 cm long, scaly. Calyx 5 - 7 mm, lobes erect, lepidote. Corolla campanulate, 1.8 - 2.8 cm long, creamy to deep yellow, lepidote and sometimes slightly pubescent outside, stamens 10; filaments pubescent at the base. Ovary lepidote, tapering into the long, slender, straight style. Capsule lepidote, 0.8 - 1.1 cm.

Distribution and habitat : India : Arunachal Pradesh (Delei valley, Mishmee Hills, Lohit district). N.E. Burma, China (S.E. Xiznag, N.W. & C. Yunnan) extending to N. Vietnam. The species seems to be rather rare in India distributed only in the eastern most part of Arunachal Pradesh bordering N.E. Burma and S.E. Xizang. In mixed forests, forest margins and alpine grassy meadows in the altitudinal range of 1500 - 4000 m. **Map 12.**

Notes: The golden yellow flowers, erect calyx lobes, silvery brown under surface of the leaves with sunken scales are the distinguishing characters of the species in the field.

The species name *xanthostephanum* means 'yellow garland', referring to its yellow flowers.

Subsection Trichoclada

Subsection **Trichoclada** (Balf.f.) Cullen, Notes R.B.G. Edinb. 36:115 (1978) and Notes R.B.G. Edinb. 39(1) : 151(1980). Cox & Cox, The Encyclopedia of Rhododendron Species 331 (1997).

Syn.: Series *Trichocladum* Balf.f., Trans. Bot. Soc. Edinb. 27:80 – 88(1916).

Subgenus *Psuedazalea* Sleumer, Bot. Jahrb. 74:525 (1949).

Type species: *R. trichocladum* Franchet

In India this subsection is represented by *R. mekongense* var. *rubrolineatum.*

R. mekongense Franchet, Journ. de Bot. 12 : 263 (1898) var. **rubrolineatum** (Balf.f. & Forrest) Cullen, Notes R.B.G. Edinb. 36:115 (1978) and op.cit. 39(1) : 155(1980). Cox & Cox, The Encyclopedia of Rhododendron Species 333 (1997).

Syn.: *R. rubrolineatum* Balf.f. & Forrest, Notes R.B.G. Edinb. 12:160 (1920).

Deciduous or semi-evergreen shrub, to 2.5 m high. Branchlets lepidote, bristly. Leaves 2.5 - 5.5 x 1.4 - 2.1 cm, obovate or rarely obovate-elliptic, rounded at apex, cuneate at base, upper surface glabrous and almost without bristles (loriform-setae), lower surface with unequal scales. Inflorescence usually 3-flowered; pedicels bristly, 1.5 - 2.2 cm long. Calyx obscurely lobed, lobes up to 2.5 mm, lepidote and bristly. Corolla widely funnel-campanulate, 1.4 - 2.3 cm long, yellow flushed with red, creamy-yellow flushed with rose outside. Stamens 10. Ovary lepidote; style impressed. Capsule lepidote.

Distribution and habitat: India : Mishmee Hills in Arunachal Pradesh bordering S.E. Xizang. China (N.W. Yunnan, S. & S.E. Xizang). Rare. In forest margins, alpine scrub, and on rocky slopes between 3,400 - 4,300 m altitude. **Map 12.**

Notes: The variety has flowers flushed with red-lines from which it derives its name.

Subsection Triflora

Subsection **Triflora** (Hutchinson) Sleumer, Bot. Jahrb. 74:536 (1949). Cullen, Notes R.B.G. Edinb. 39(1): 61(1980). Cox & Cox, The Encyclopedia of Rhododendron Species 335 (1997).

Syn.: Series *Triflorum* sensu Hutchinson in Stevenson (ed.) The Species of Rhododendron 758(1930) pro parte ; sensu Davidian, R.H.S. Rhodo. Yearbook 17:156 (1963) pro parte.

Though a large and variable subsection, it is represented in India only by *R. triflorum* Hook.f.

Type species: *R.triflorum* Hook.f.

R. triflorum Hook.f., Rhododendrons of Sikkim Himalaya t. 19 (1849). Cullen, Notes R.B.G. Edinb. 39(1): 77(1980). Cox & Cox, The Encyclopedia of Rhododendron Species 345(1997).

Shrub, 5 (- 7) m tall, often thinly-clothed with leaves. Bark smooth, reddish-brown, peeling. Leaves usually evergreen, sometimes semi-deciduous, aromatic, 3 - 7 x 1.3 - 2 cm, oblong-lanceolate to elliptic, dark green and nearly glabrous on the upper surface, glaucous or pale green on the lower surface with brown, very small, rimless scales. Inflorescence terminal, 2 - 4-flowered; pedicels 8 - 13 mm long, scaly. Calyx small, inconspicuous, rarely 5-lobed, scaly. Corolla widely-funnel shaped, zygomorphic, 2 - 3.3 cm long, 2 - 4.3 cm wide, pale or lemon-yellow, spotted with red or pink; tube 7 - 12 mm, densely lepidote and pubescent outside. Stamens 10, filaments pubescent at the base. Ovary lepidote; style glabrous or rarely pubescent at the base. Capsule lepidote, 10 - 13 mm long, cylindric.

Based on the corolla shape the species is divided into two varieties:

1. Corolla widely funnel-shaped, 2.3 - 3.6 cm across .... var. *triflorum*

1a. Corolla saucer-shaped, flat, 3.8 - 4.3 cm across .... var. *bauhiniiflorum*

R. triflorum Hook.f. var. **bauhiniiflorum** (Watt ex Hutchinson) Cullen, Notes R.B.G. Edinb. 36:109 (1978).

Syn.: *R. bauhiniiflorum* Watt ex Hutchinson in Stevenson (ed.) The Species of Rhododendron 785 (1930).

Corolla saucer-shaped, 3.8 - 4.3 cm across, greenish to clear-yellow.

Distribution and habitat: India : Nagaland and Manipur. Endemic. Less frequent, in the forested hill slopes in the altitudinal range of 2400 - 2900 m. **Map 12.**

R.triflorum Hook.f.var. **triflorum. (Photo Nos. 39, 40).**

Syn.: *R. deflexum* Griffith, Notulae 4:303 & t. 519 (1854). *R. triflorum* var. *mahogani* Hutchinson, Gard. Chron. 101:135 (1937).

Distribution and habitat: India: Sikkim, Arunachal Pradesh. E. Nepal, Bhutan, S.E. Xizang, Burma-Tibet Frontier. In varied habitats in the altitudinal range of 2100 - 4000 m. **Map 12.**

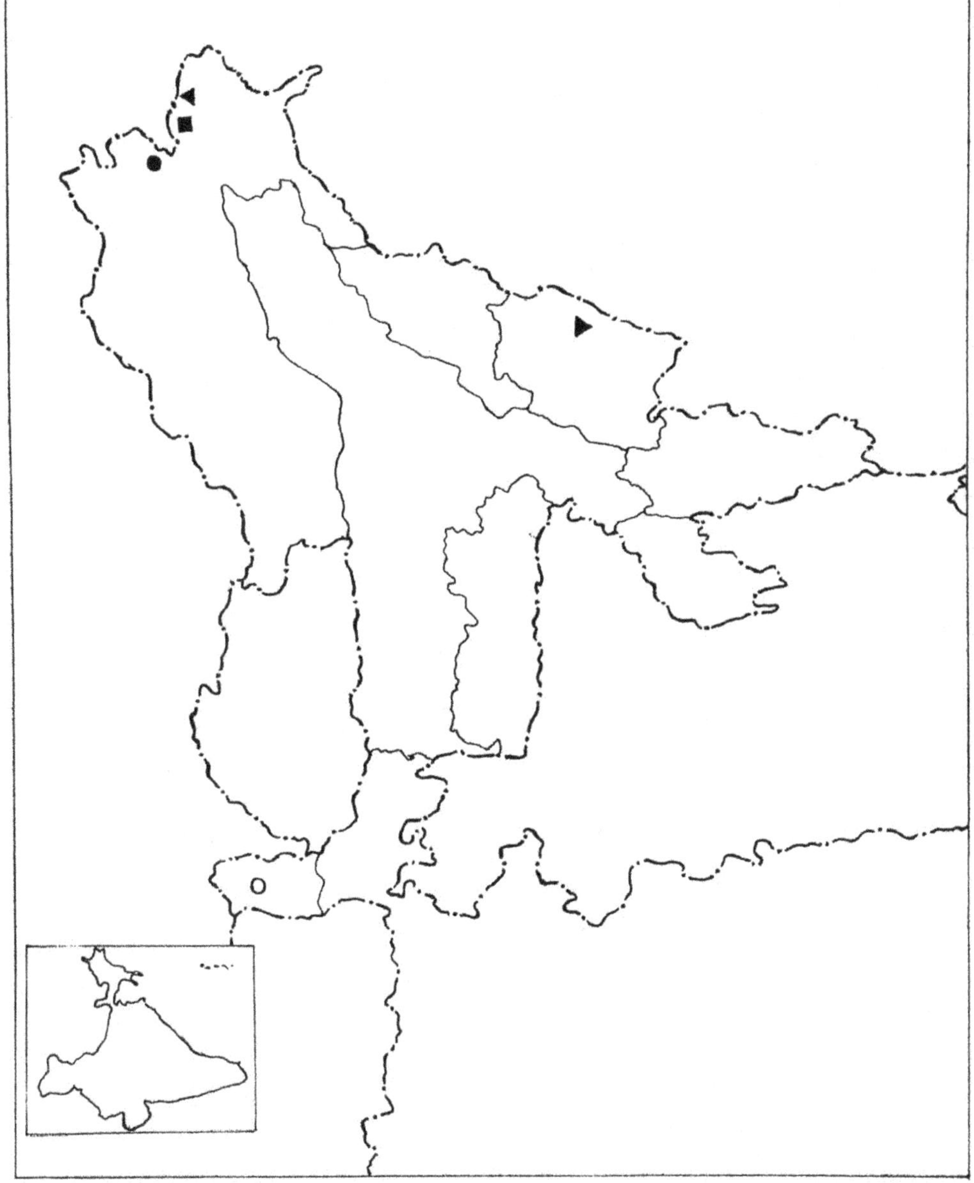

Map 12. ■ R.tephropeplum. ● R.xanthostephanum. ▲ R.mekongense var.rubrolineatum. ▼ R.triflorum var.bauhiniiflorum. O var.triflorum.

(Maps are only indicative and not to scale)

Subsection Uniflora

Subsection **Uniflora** (Cowan & Davidian) Sleumer, Bot. Jahrb. 74:532 (1949). Cullen, Notes R.B.G. Edinb. 39(1): 120(1980). Cox & Cox, The Encyclopedia of Rhododendron Species 354(1997).

Syn.: Series *Lepidotum* sensu Hutchinson in Stevenson (ed.) The Species of Rhododendron 437 (1930) pro parte.

Series *Uniflora* sensu Cowan & Davidian, R.H.S. Rhodo. Yearbook 3:101 (1948).

Low shrubs, often prostrate, stoloniferous, or mat-forming, 0.07 - 0.9 m high. Leaves evergreen, to 3 cm long, margins sometimes revolute or crenulate, scaly on the lower surface. Inflorescence terminal, 1 - 3-flowered; pedicels rigid, erect and accrescent in fruit. Calyx 1 - 7 mm. Corolla funnel-campanulate, 5-lobed, densely pilose outside, pink, purple or yellow. Stamens 10, declinate, filaments pubescent towards base. Ovary lepidote; style impressed, slender and straight. Capsule lepidote. Seeds unwinged and obscurely finned.

Type species: *R. uniflorum* Kingdon-Ward

Notes: The species in this subsection are characterized by their low habit and densely pilose corolla.

Key to the species of subsection Uniflora

1. Leaf margins notched .... *ludlowii*

1a. Leaf margins entire 2

2. Corolla campanulate, 0.8 - 2.1 cm, tube 0.7 - 1.4 cm; style shorter than the stamens .... *pumilum*

2a. Corolla widely funnel-shaped, 2.3 - 3.2 cm, tube 1.2 - 1.8 cm, style longer than the stamens .... *imperator*

R. imperator Kingdon-Ward, Gard. Chron. 88:299 (1930). Cox & Cox, The Encyclopedia of Rhododendron Species 354(1997).

Syn.: *R. patulum* Kingdon-Ward, Gard. Chron. 88:298 (1930).

R. uniflorum Kingdon-Ward var. *imperator* (Kingdon Ward) Cullen, Notes R.B.G. Edinb. 36:113 (1978).

Prostrate shrub, *ca* 0.3 m high. Leaves semi-deciduous, 1.8 - 3.8 x 0.4 - 1 cm, lanceolate or oblanceolate, very pointed at both ends, upper surface dark and shiny, lower surface pale green, with brown scales, 1 - 6 x their own diameter apart. Inflorescence 1 - 2-flowered; pedicels 0.9 - 1.2 cm in flower, up to 2.5 cm in fruit, lepidote. Calyx lepidote, reddish. Corolla widely funnel-shaped, 2.3 - 3.2 cm long, bright to dark pinkish-purple, sometimes with deep crimson spots, pubescent outside. Stamens 10. Ovary lepidote; style impressed, longer than stamens, slender, straight.

Distribution and habitat: India : Arunachal Pradesh (Delei valley, Mishmee Hills – Tibet-Assam frontier). China (S.E. Xizang), N. Burma. In temperate shaded gullies, on rocks and steep slopes in between 2500 - 3500 m. **Map 13.**

Notes: Cox & Cox (*The Encyclopedia of Rhododendron Species* 354: 1997) consider that the more widely distributed *R. imperator* with large typical populations having narrow lanceolate or oblanceolate leaves pointed at both the ends and comparatively larger flowers to be distinctive from *R. uniflorum* rather than a mere variety of it. They also state that the Type specimen of *R. patulum* from Delei valley, Mishmee Hills, Arunachal Pradesh (*Kingdon-Ward* 8260, collected on 30.5.1928) is referable to *R. imperator.* They further state that plants of *R. platulum* in cultivation are usually of *R. pemakoense,* which is not reported in India so far.

R. ludlowii Cowan, Notes R.B.G. Edinb. 19:243(1937). Cox & Cox, The Encyclopedia of Rhododendron Species 355(1997).

Small shrub, up to 0.3 m high, usually spreading or creeping. Branchlets lepidote. Leaves 1.2 - 1.6 x 0.8 - 1 cm, broadly obovate or oblong-obovate, obtuse at apex, rounded at base, margins crenate, upper surface dark green, lower surface pale green with large brown distantly placed scales and prominent venation. Inflorescence 1 - 2-flowered; pedicels 1.5 - 2 cm in flower, lepidote. Calyx large, 5 - 7 mm, lobes ciliate. Corolla broadly-campanulate, 1.5 - 2.5 cm long, yellow with reddish-brown spots, densely pubescent and scaly outside. Stamens 10, filaments pubescent at base. Ovary lepidote; style impressed, glabrous, longer than the stamens and shorter than corolla.

Distribution and habitat: India :Arunachal Pradesh. China (S. Xizang). Creeping amongst moss on open rocky hill slopes in the altitudinal range of 4,000 - 4,300 m. **Map 13.**

Notes: The species is included on the authority of Mr. P.A. Cox (pers. comm.. to A.R.K.Sastry). The species is named after F. Ludlow, who has extensively collected plants in Bhutan and S. Xizang along with Sheriff and Hicks.

R. pumilum Hook.f., Rhododendrons of Sikkim Himalaya t.14 (1849).

Prostrate shrublet, to 10 cm high. Young branchlets lepidote and pubescent. Leaves 0.8 - 1.9 x 0.4 - 1.1 cm, elliptic to broadly elliptic, apex rounded or acute, base cuneate, upper surface dark green, lower surface glaucous and scaly, scales distant, golden yellow. Inflorescence 1 - 3-flowered; pedicels 1 - 2 cm in flower, elongating to 4 - 6 cm in fruit.

Calyx reddish, deeply lobed, lobes 2 - 3.5 mm long, ovate-oblong or oblong, scaly. Corolla 0.8 - 2.1 cm long, campanulate, mouth slightly oblique, pink to purple, densely pilose, sparsely scaly mostly on the lobes. Stamens 10, filaments pubescent below. Ovary densely scaly; style impressed, longer or equaling the stamens, included in the corolla. Capsule 0.7 - 1.0 cm, scaly.

Distribution and habitat: India : Sikkim, Arunachal Pradesh. E. Nepal, Bhutan, N.E. Burma, China (S. & S.E. Xizang) between 3,400 - 4,300 m. Frequent, in open situations along the mountain slopes, on mossy rock borders and stream banks. **Map 13.**

Notes: The species epithet *pumilum* refers to its dwarfish habit.

Subsection Virgata

Subsection **Virgata** (Hutchinson) Cullen, Notes R.B.G. Edinb. 36:113 (1978). Cox & Cox, The Encyclopedia of Rhododendron Species 358(1997).

Syn.: Series *Virgatum* Hutchinson in Stevenson (ed.) The Species of Rhododendron 827 (1930) pro max. parte. Subgenus *Pseudorhodorastrum* Sleumer section *Rhabdorhodion* Sleumer Bot. Jahrb. 74:529 (1949).

A monotypic subsection easily distinguished from other lepidote rhododendrons by the lateral, 1-flowered inflorescence and tailed seeds.

Type species: *R. virgatum* Hook.f.

R. virgatum Hook.f., Rhododenrons of Sikkim Hiamalaya, t. 26 (1849). subsp. **virgatum.** Cox & Cox, The Encyclopedia of Rhododendron Species 358 (1997). **(Photo No. 41).**

Shrub, 0.3 - 2.4 m high, rather loosely branched; terminal bud always vegetative. Young growth lepidote. Leaves 1.8 - 8 x 0.5 - 2 cm, recurved, broadly lanceolate or oblong-ovate,upper surface dark green, lower surface pale with unequal brown scales. Inflorescence 1 (- 2)-flowered, axillary. Calyx 2 - 3 mm, 5-lobed, lobes rounded, sometimes fringed with filiform-acicular hairs. Corolla tubular-campanulate, 2.5 - 3.9 cm long, pale to deep pink or mauve coloured, tube 1.1 - 2 cm long. Stamens 10, filaments pubescent near the base. Ovary densely lepidote; style straight, longer than stamens, pilose or lepidote at the base. Capsule lepidote, 0.9 - 1.3 cm long. Seeds tailed at both ends.

Distribution and habitat : India : Sikkim, Arunachal Pradesh. Bhutan, China (S.& S.E. Xizang). Frequent along forest margins in alpine scrub, stony slopes and dry banks in the altitudinal range of 2400 - 3800 m (8000 - 12000 ft). **Map 13.**

Notes: A distinct species with single or rarely two-flowered axillary inflorescences, funnel-shaped corolla, lobed calyx and the willowy twigs by which it gets the specific epithet *virgatum.*

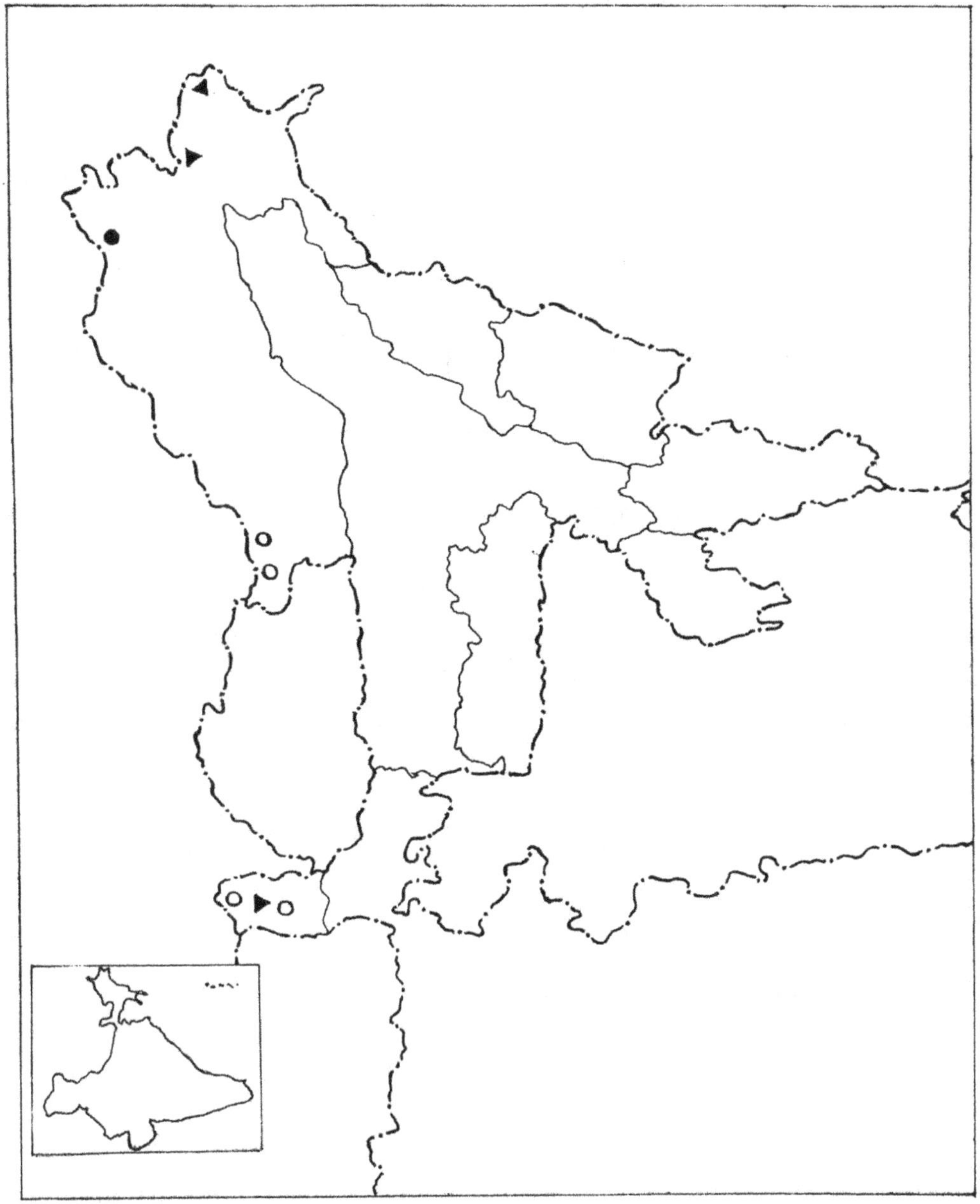

Map 13. ▲ R.imperator. ● R.ludlowii. ▼ R.pumilum. O R.virgatum ssp.virgatum.

(Maps are only indicative and not to scale)

5
Subgenus Hymenanthes

Subgenus **Hymenanthes** (Blume) K. Koch, Dendrologie 2:170 (1872). Chamberlain, Notes R.B.G. Edinb. 39(2) : 219(1982). Cox & Cox, The Encyclopedia of Rhododendron Species 9(1997).

Syn.: Genus *Hymenanthes* Blume, Bijdragen Fl. Nederlandsch Indie 862 (1826).

Subgenus *Eurhododendron* auct. plur, including Sleumer (1949), non K. Koch (1872).

Type species: *Rhododendron japonicum* (Blume) Schneider (*Hymenanthes japonica* Blume).

Description of the subgenus **Hymenanthes** as for sect. **Ponticum** being the only section.

Section **Ponticum** G. Don, Gen. Hist. 3:843 (1834).

Syn.: Section *Leiorrhodium* Rehder, J. Arnold Arbor. 15:269 (1934).

Shrubs or large trees. Leaves evergreen. Scales absent. Plants glabrous or with indumentum on the young shoots, leafbud scales, leaves, pedicels, calyx, ovary, style and stamens. Calyx obsolete or well-developed. Corolla 5 – 10-lobed, open – or ventricose-campanulate to tubular-campanulate, with or without nectar pouches. Stamens 10 – 20, declinate. Ovary 5 – 20-celled. Capsule with hard, woody valves. Seeds winged or without wings.

Type species: *R. ponticum* L. (from Pontic Range, E. Turkey).

Key to the subsections of section Ponticum

1. Calyx 4 – 20 mm 2

1a. Calyx 1 – 3 mm 3

2. Leaves underneath glabrous or with vestiges of stellate hairs specially on the midrib; calyx small, *ca* 4 mm, glandular or sometimes with a few hairs 4

2a. Leaves underneath with stipitate glands or bristles or dense ramiform hairs or strongly papillate with a thin dendroid indumentum or with red punctate hair-bases which appear as small red flecks; calyx cup-shaped, 6 - 20 mm or split to 10 mm, glandular or eglandular 5

3. Ovary densely tomentose 6

3a. Ovary glabrous 7

4. Leaves underneath with stellate hairs specially on the mid-rib or nearly glabrous; calyx 3 - 4 mm; ovary and style tomentose and glandular .... *Parishia*

4a. Leaves underneath glabrous; calyx 10 - 20 mm, saucer-shaped or irregular; ovary and style glandular-pubescent .... *Fortunea*

5. Leaves underneath with stipitate glands or bristles or dense ramiform hairs; calyx split to 10 mm long, stipitate glandular .... *Glischra*

5a. Leaves underneath usually strongly papillate or with thin dendroid hairs or with red punctate hair-bases which appear as small red flecks; calyx 6 - 20 mm, glandular or eglandular... *Thomsonia*

6. Leaves underneath with a dense, unistrate, light brown to rufous, lanate or crisped indumentum composed of dendroid hairs .... *Lanata*

6a. Leaves underneath not as above 8

7. Leaves underneath with a dense unistrate or bistrate, granular, woolly or silky tomentum or fulvous lanate tomentum consisting of fasciculate hairs; ovary glabrous 15

7a. Leaves underneath usually with a dense fawun-rusty, brown or loose, scattered indumentum, or nearly glabrous or woolly with a thin veil of tomentum or with slightly granular, glandular felted, greyish-brown to orange-brown indumentum; ovary glabrous or slightly glandular or red-brown tomentose 11

8. Leaves underneath glabrous or with a compacted, continuous or patchy , whitish or buff-rufous indumentum........................ *Neriiflora*

8a. Leaves underneath with a dense, plastered or white to rufous woolly indumentum often with cup-shaped hairs 9

9. Leaves underneath with a dense, spongy to plastered indumentum 10

9a. Leaves underneath with a white to rufous, woolly indumentum often with cup-shaped hairs .... *Falconera*

10. Leaves 7 - 19 cm long, 1.5 - 6.5 cm broad .... *Arborea*

10a. Leaves 15 - 70 (-90) cm long, 8 - 30 cm broad .... *Grandia*

11. Leaves underneath with a dense fawn-rusty brown or loose scattered indumentum or nearly glabrous; ovary and style glabrous .... *Campanulata*

11a. Leaves underneath pale glaucous or nearly glabrous with slightly granular, glandular, brown to orange-brown or somewhat with a dense, compacted indumentum of radiate hairs, felted greyish or usually tomentose or with a thin veil of tomentum; ovary and style glabrous or glandular or densely brown tomentose 12

12. Leaves underneath pale glaucous or nearly glabrous or with a thin woolly indumentum; ovary glabrous or tomentose or glandular; style glabrous or glandular at base to half its length 13

12a. Leaves underneath glabrous or with a slightly granular, felted, greyish-brown or somewhat dense compacted indumentum consisting of radiate hairs or with a thin veil of indumentum; ovary and style glabrous or glandular or densely brown tomentose.. 14

13. Leaves underneath glabrous or with a woolly indumentum; ovary glabrous or tomentose to glandular; style glabrous; nectar pouches present .... *Barbata*

13a. Leaves underneath pale glaucous or nearly glabrous; ovary stipitate glandular; style glandular at base to half its length or eglandular; nectar pouches absent....................................... *Campylocarpa*

14. Leaves underneath glabrous or with a thin veil of indumentum; ovary and style glabrous or glandular .... *Irrorata*

14a. Leaves underneath with a slightly granular, felted, greyish-brown to orange-brown indumentum or somewhat compacted indumentum consisting of radiate hairs; ovary glabrous or densely red-brown tomentose, style glabrous .... *Taliensia*

15. Leaves underneath with a dense, unistrate or bistrate, granular, woolly or silky indumentum; ovary glabrous .... *Fulva*

15a. Leaves underneath with a unistrate, dense, fulvous, lanate – tomentose indumentum consisting of fasciculate hairs; ovary glabrous .... *Fulgensia*

Subsection Arborea

Subsection **Arborea** Sleumer, Bot. Jahrb. 74:54 (1949). Chamberlain, Notes R.B.G. Edinb. 39(2) : 328(1982). Cox & Cox, The Encyclopedia of Rhododendron Species 9(1997).

Syn.: Series *Arboreum* subseries *Arboreum* sensu Tagg in Stevenson (ed.) The Species of Rhododendron 12 (1930).

Trees, up to 30 m high with well-developed columnar trunks. Bark rough, fissured. Young branches densely tomentose. Leaves thick, leathery, elliptic to oblanceolate, persistent for 1 - 3 years, upper surface dark green, glabrous on maturity, lower surface with a dense spongy to plastered, white to yellowish-brown tomentum of unistrate or bistrate or dendroid hairs, sometimes with a floccose rufous upper layer. Inflorescence dense, 10 - 35-flowered. Calyx minute. Corolla 5-lobed, campanulate or tubular-campanulate with nectar pouches, red, pink, white, mauve coloured. Stamens 10. Ovary densely tomentose, sometimes glandular; style glabrous.

Type species : *R. arboreum* Smith

Key to the species of subsection Arborea

1. Leaves 16 - 22 cm long, indumentum bistrate, with a lanate-tomentose upper layer and a compacted lower layer; inflorescence 20 - 35-flowerd .... *lanigerum*

1a. Leaves 6.5 - 19 cm long, if more than 15 cm then indumentum more or less unistrate and compacted; inflorescence 10 - 20-flowered 2

2. Corolla fleshy, 3 - 5 cm long, varying shades of rose-pink to deep carmine, rarely even white; a regular tree with well-defined trunk branching from above breast height .... *arboreum*

2a. Corolla not so fleshy, 3 - 3.5 cm long, deep lilac to deep magenta; a smaller tree with trunk branching from near the base .... *niveum*

R. arboreum Smith, Exot. Bot. 1:9 t. 6 (1805). C.B. Clarke in Hook.f., Fl. Brit. India 3:465 (1882). Chamberlain, Notes R.B.G. Edinb. 39(2):328(1982). Cox & Cox, The Encyclopedia of Rhododendron Species 9(1997).

Tree, up to 30 m high (rarely to 50 m), trunk well-developed. Bark rough, fissured, dark brown. Young growth densely tomentose. Leaves 6 - 20 x 2 - 5 cm, oblong-lanceolate to oblong-oblanceolate, thick and leathery,upper surface dark green and glossy, lower surface silvery white to fawn to cinnamon or rusty-brown indumentum with dendroid hairs, sometimes with a floccose rufous upper layer; petioles 1 - 2 cm, loosely tomentose, inter mixed with glands, or glabrescent at maturity. Inflorescence 10 - 20-flowered, dense; rachis 1.5 - 2 cm long; pedicels 5 - 10 mm, pilose and glandular. Calyx 1 - 3 mm, lobes rounded, sparsely glandular, hairy or nearly glabrous. Corolla fleshy, tubular-campanulate, 3 - 5 cm long, pink to deep crimson to scarlet, rarely white, sometimes spotted, nectar pouches present. Ovary white-tomentose or glandular. Capsule 1.5 - 3 cm long.

Notes : *R. arboreum* is a wide spread and an extremely variable species, recognized into many subspecies and varieties. The species derives its name from its 'tree-like' habit.

Key to the subspecies and varieties of *R. arboreum*

1. Leaves underneath with a compacted indumentum 2

1a. Leaves underneath with a spongy or floccose brown to dark brown indumentum 3

2. Leaves 10 - 19 cm long, lower surface with white to silvery indumentum .... subsp. *arboreum*

2a. Leaves 6.5 - 11 cm long, lower surface with yellowish-brown indumentum .... subsp. *cinnamomeum* var. *roseum*

3.	Leaf indumentum on the lower surface bistrate, the upper layer loose and the lower layer whitish to yellowish brown hairy	subsp. *cinnamomeum* var. *cinnamomeum*
3a.	Leaf indumentum on the lower surface unistrate, white to fawn, more or less spongy	4
4.	Leaves rounded at apex, bullate	subsp. *nilagiricum*
4a.	Leaves acute at apex, reticulate	5
5.	Leaves 5 - 16 x 2 - 3 cm	subsp. *delavayi* var. *delavayi*
5a.	Leaves 7.5 - 15 x 1.8 - 3 cm	subsp. *delavayi* var. *peramoenum*

R. arboreum Smith subsp. **arboreum. (Photo Nos. 42, 43).**

Leaves 10-19 x 3 - 5 cm; apex acute, reticulate above, with a compacted, white to silvery indumentum below. Corolla bright red to carmine, raely pink or white.

Distribution and habitat: India : Himalayan foothills from Kashmir to Sikkim. Nepal, Bhutan. Common, in mixed forests in the altitudinal range of 1,850 - 3,000 m. **Map 14.**

R. arboreum Smith subsp. **cinnamomeum** (Lindley) Tagg in Stevenson (ed.) The Species of Rhododendron 17 (1930). var. **cinnamomeum** (Wall. ex G.Don) Lindley, Edwards Bot. Reg. 23: t. 1982(1837). Chamberlain, Notes R.B.G. Edinb. 39(2) : 330(1982). Cox & Cox, The Encyclopedia of Rhododendron Species 10(1997). **(Photo Nos. 44, 45).**

Syn.: *R. cinnamomeum* Wall. ex G.Don, Gen. Syst. 3 : 844(1834).

R. campbelliae Hook.f., Rhododendrons of Sikkim Himalaya t.6 (1849).

Leaves 6.5 - 11 x 2.5 - 6 cm, 2.5 - 3.7 x as long as broad, apex acute, upper surface reticulate, lower surface with bistrate indumentum, the upper layer loose and floccose, rufous, the lower layer whitish to fawn and compacted. Corolla pink to carmine, sometimes white.

Distribution and habitat: India : Sikkim, Darjeeling Hills in West Bengal. E. Nepal. Fairly frequent in open forests and rocky slopes, in the altitudinal range of 2,750 - 3,650 m. **Map 14.**

Notes: The cinnamon-coloured indumentum on the lower surface of the leaves signifies its identity and name.

R. arboreum Smith subsp. **cinnamomeum** (Lindley) Tagg var. **roseum** Lindley, Bot. Reg. 15: t. 1240 (1829). Chamberlain, Notes R.B.G. Edinb. 39(2) : 331(1982). Cox & Cox, The Encyclopedia of Rhododendron Species 11(1997).

Syn.: *R.arboreum* Smith var.*album* Wallich, Pl.Asiat.Rar.2:23,t.123(1830).

Leaves 6.5 - 11 x 2.2 - 4.5 cm, 2.7 - 3.5 x as long as broad, apex acute, upper surface reticulate, lower surface with a light fawn, sometimes whitish, compacted unistrate indumentum. Corolla pink to carmine, rarely white.

Notes: Cox & Cox state that var. *roseum* intergrades with subsp. *arboreum* and subsp. *delavayi* and that many natural hybrids of this variety have been found, crossed with *R. campanulatum, R. wallichii* and *R. barbatum* etc. The var. *roseum* includes white-flowered forms, formerly treated as var. *album* which have bisrate indumentum and white corolla with purple spottings in the corolla throat.The white flowered forms earlier known from Nepal are now reported from Darjeeling hills at Kalipokhri.

Distribution and habitat: India : Sikkim, Darjeeling Hills in West Bengal, Arunachal Pradesh. E. Nepal, Bhutan, China (S. Xizang). In open forests and rocky slopes in between 2,400 - 4,000 m altitude. **Map 14.**

R. arboreum Smith subsp. **delavayi** (Franchet) Chamberlain, Notes R.B.G. Edinb. 37:328(1979) var. **delavayi.** Cox & Cox, The Encyclopedia of Rhododendron Species 12(1997).

Syn.: *R. delavayi* Franchet, Bull. Soc. Bot. France 33:231 (1886).

Leaves relatively broad, 5 - 16 x 2 - 2.4 cm, 2.8 - 4 x as long as broad, apex acute, upper surface reticulate, lower surface with a unistrate, spongy, brown or fawn, rarely white indumentum. Corolla deep crimson to carmine.

Distribution and habitat: India : N.E. India: Meghalaya, Manipur, Assam, Arunachal Pradesh. Burma, China (Yunnan), Thailand. Frequent, in mixed forests and in conifer forests, in between 1,500 - 3,200 m altitude. **Map 15.**

Notes: The species is named after L'Abbe Jean M. Delavay, a plant collector in W. China.

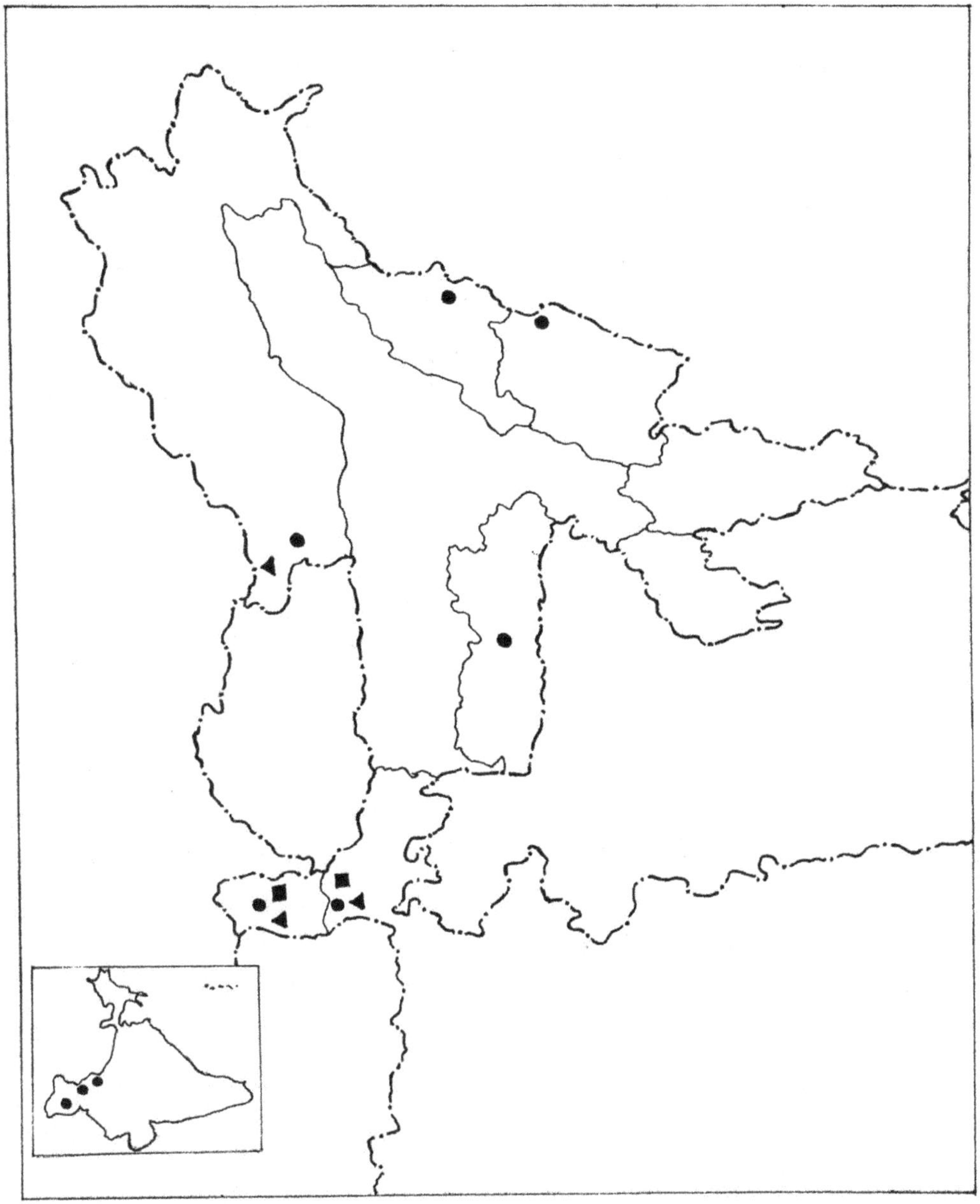

Map 14. ● R.arboreum ssp.arboreum. ■ ssp.cinnamomeum var.cinnamomeum. ▲ ssp.cinnamomeum var.roseum.

(Maps are only indicative and not to scale)

R. arboreum Smith subsp. **delavayi** (Franchet) Chamberlain var. **peramoenum** (Balf.f. & Forrest) Chamberlain, Notes R.B.G. Edinb. 37:328 (1979). Cox & Cox, The Encyclopedia of Rhododendron Species 12(1997).

Syn.: *R. peramoenum* Balf.f. & Forrest, Notes R.B.G. Edinb. 13:56 (1920).

Leaves relatively narrow (than in the var. *delavayi*), 7.5 - 15 x 1.8 - 3 cm, 4.5 - 6.5 x as long as broad, apex acute to shortly cuspidate, lower surface with a loose spongy, fawn or brown indumentum. Corolla usually crimson to carmine.

Distribution and habitat: India: Subansiri District, Arunachal Pradesh. China (W. Yunnan). Frequent, in mixed temperate forests associated with other *Rhododendron* species and Magnoliaceae members, in between 2,440 - 3,350 m altitude. **Map 15.**

Notes: This variety has been first collected by one of the authors (A.R.K.S) in Subansiri district in April, 1965 and has been reported as a new distribution record for India under the name *R. peramoenum.*

The epithet *peramoenum* means 'very pleasing'.

R. arboreum is a parent of several clones that are in cultivation. Its subspecies *cinnamomeum* is very commonly cultivated in U.K.

R. arboreum Smith subsp. **nilagiricum** (Zenker) Tagg in Stevenson (ed.) The Species of Rhododendron 15 (1930). Chamberlain, Notes R.B.G. Edinb. 39(2) : 331(1982). Cox & Cox, The Encyclopedia of Rhododendron Species 13(1997).

Syn.: *R. nilagiricum* Zenker, Amer. Sci. Nat. ser. 2, 6:150 (1836).

Leaves 8.5 - 12 x 3.8 - 6 cm, 1.8 x 2.4 x as long as broad, apex more or less rounded, apiculate, upper surface rugose or bullate, lower surface with a spongy yellowish-brown indumentum. Corolla deep crimson, crimson-rose, occasionally pink.

Distribution and habitat : India : Nilagiris and other South Indian hill ranges in Tamil Nadu & Kerala. Endemic. In mixed forests, in between 1,800 - 2,250 m alt. **Map 15.**

Notes: The taxon derives its name from its distribution in the Nilagiri mountains.

R. lanigerum Tagg, Notes R.B.G. Edinb. 16:199 (1931). Chamberlain, Notes R.B.G. Edinb. 39(2) : 332(1982). Cox & Cox, The Encyclopedia of Rhododendron Species 13 (1997).

Syn.: *R. silvaticum* Cowan, Notes R.B.G. Edinb. 19:185 (1936).

Shrub or a small tree, 2.7 - 6 m high. Branchlets tomentose at first. Bark flacking, greenish or greyish or brownish. Flower buds large, round. Leaves 10 - 22 x 5 - 7.5 cm, elliptic to oblanceolate, upper surface rugulose, becoming glabrous at maturity with traces of indumentum along the midrib, lower surface with a whitish to fawn dense fine woolly bistrate indumentum; petioles 1.5 - 2 cm, velutinous. Inflorescence 20 - 35-flowered, dense; pedicels *ca* 10 mm, glabrous. Calyx fleshy, *ca* 2 mm, with minute teeth, glabrous. Corolla campanulate, 4 - 5 cm long, deep pink to reddish-purple, cherry red to scarlet crimson, with darker nectar pouches. Ovary densely tomentose, style glabrous. Capsule short, cylindrical, *ca* 20 x 7 mm.

Distribution and habitat : India : Delai valley – Eastern Arunachal Pradesh. China (S. Xizang). Rare; on ridges and amongst fir and bamboo, in between 2,600 - 3,400 m altitude **Map 15.**

Notes: The species name *lanigerum* refers to the woolly (tomentose) nature of the branchlets.

R. niveum Hook.f., Rhododendrons of Sikkim Himalaya t. 4 (1851). Chamberlain, Notes R.B.G. Edinb. 39(2) : 332(1982). Cox & Cox, The Encyclopedia of Rhododendron Species 14 (1997). **(Photo Nos. 46, 47).**

Bushy shrub or small tree, 2.5- 6 m high. Branchlets densely white tomentose. Bark pale brown to greyish. Leaves 10 - 16.5 x 4 - 6.3 cm, oblanceolate to elliptic, upper surface dark green, glabrous, lower surface with a dense bistrate woolly indumentum, white at first, turning to grey to fawn on maturity; petioles 1 - 1.5 cm, floccose. Inflorescence 15 - 20-flowered, dense; pedicels *ca* 10 mm, densely white to rufous-tomentose. Calyx 1 - 2 mm, minute. Corolla tubular-campanulate, 4 - 5 cm long, deep magenta to deep lilac with darker nectar pouches. Ovary densely white to fawn-tomentose; style glabrous. Capsule shortly cylindrical, *ca* 20 x 8 mm.

Distribution and habitat : India : Sikkim. Bhutan. Rare; in rocky valleys and mixed temperate forests, in between 3,000 - 3,700 m altitude **Map 15.**

Notes: The species name *niveum* means 'snow-like', and refers to the dense white woolly indumentum that covers the young shoots. This species is one of the most distinctive amongst all rhododendrons in its deep magenta to deep lilac flowers with darker nectar pouches.

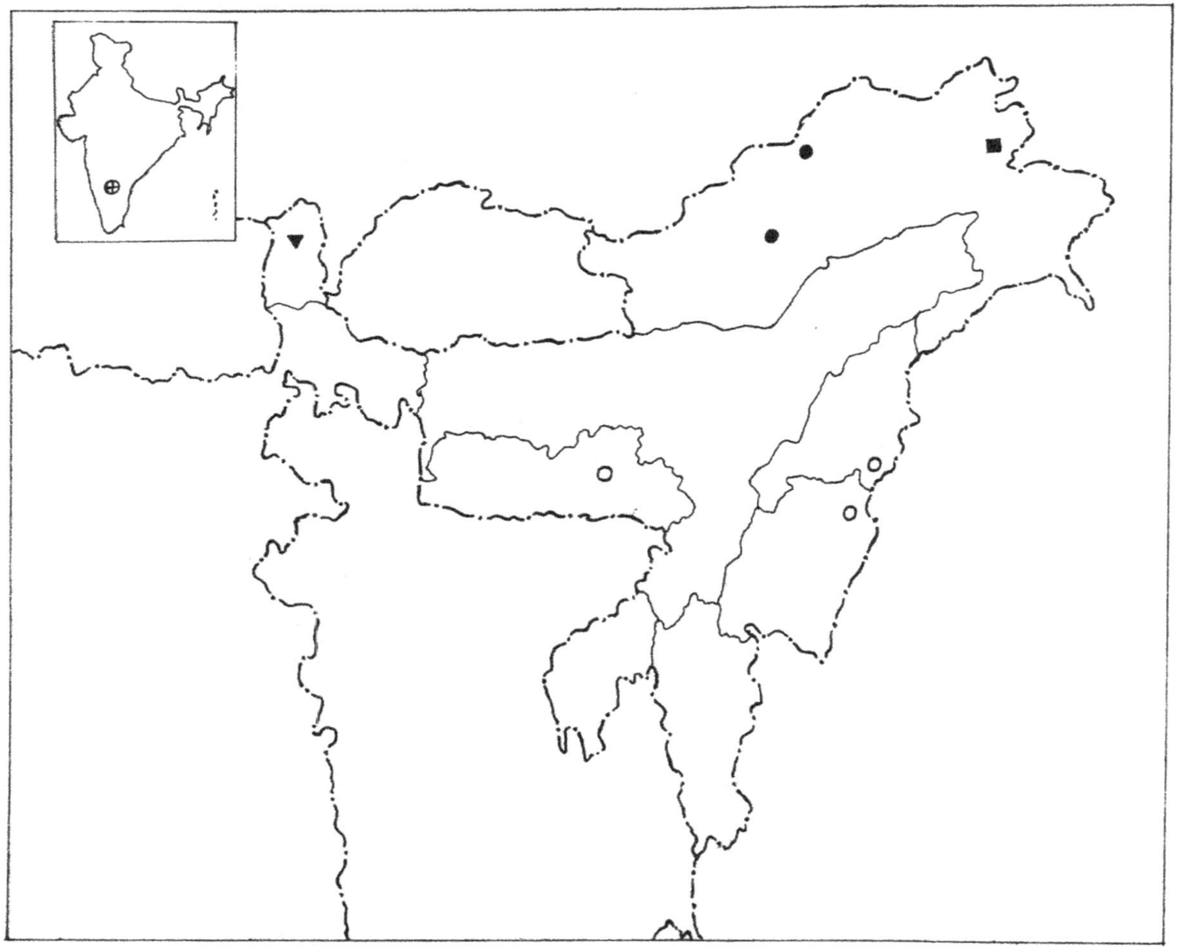

Map 15. ○ R.arboreum ssp.delavayi var.delavayi. ● var.peramoenum. ⊕ R.arboreum ssp.nilagaricum. ■ R.lanigerum. ▼ R.niveum

(Maps are only indicative and not to scale)

Photo No. 33. *R.dalhousiae* var.*dalhousiae*.

Photo No. 34. *R.dalhousiae* var.*rhabdotum*.

Photo No. 35. *R.dalhousiae* var.*rhabdotum*.

Photo No. 36. *R.lindleyi*.(note long, curved style).

Photo No. 37. *R.ciliatum*

Photo No. 37a. *Rhododendron* sp. in cultivation at Farnborough, Kent, U.K.

Photo No. 38. *R.parryae - R.walongense* affinity.

Photo No. 39. *R.triflorum* var. *triflorum*.

Photo No. 40. *R.triflorum* var. *triflorum*.

Photo No. 41. *R.virgatum* ssp.*virgatum*.

Photo No. 42. *R.arboruem* in full bloom.

Photo No. 43. *R.arboreum* with bright red flowers.

Photo No. 44. *R.arboreum* ssp.*cinnamomeum*.

Photo No. 45. *R.arboreum* ssp.*cinnamomeum*.

Photo No. 46. *R.nivuem.*

Photo No. 47. *R.nivuem.*

Subsection Barbata

Subsection **Barbata** Sleumer, Bot. Jahrb. 74:544 (1949) pro parte.

Syn.: Series *Barbatum* subseries *Barbatum* sensu Tagg in Stevenson (ed.) The Species of Rhododendron 126 (1930) pro parte. Chamberlain, Notes R.B.G. Edinb. 39(2) : 379(1982). Cox & Cox, The Encyclopedia of Rhododendron Species 25 (1997).

Shrubs or small trees. Branchlets usually bristly. Bark smooth, peeling, plum, purple to reddish. Leaves dark green, usually strongly concave, elliptic to broadly obovate, upper surface glabrous on maturity, rugulose, lower surface with a thin woolly indumentum composed of lanate dendroid hairs. Inflorescence 10 - 20-flowered, dense. Calyx minute to large and cup-shaped, up to 15 mm long, often coloured. Corolla tubular-campanulate, with nectar pouches, scarlet to crimson, rarely rose-pink. Stamens 10, filaments glabrous. Ovary glabrous to densely stipitate glandular, with or without a rufous tomentum of dendroid (tree-like) hairs; style glabrous.

Type species: *R. barbatum* Wall. ex G. Don

Key to the species of subsecton Barbata

1.	Petioles 0 - 5 mm; leaves and petioles glabrous	*succothii*
1a.	Petioles 5 - 20 mm; leaves and / or petioles setose or bristly..	2
2.	Leaves 2.3 - 3.2 x as long as broad, apex acute to acuminate, lower surface with a thin, continuous loose indumentum, white at first, turning to buff to rufous or greyish white or glabrous or with dendroid hairs	3
2a.	Leaves 1.5 - 2.2 x as long as broad, apex rounded, lower surface with stout gland-tipped setae intergrading with bristles on the midrib; corolla brick-red	*exasperatum*
3.	Lower surface of leaves usually glabrous or with scattered dendroid hairs and stipitate glands	*barbatum*

3a. Lower surface of leaves with a thin continuous, loose indumentum, white at first turning to buff to rufous or greyish white 4

4. Leaves oblong-lanceolate to elliptic-lanceolate, upper surface rugulose, lower surface with a thin, continuous, loose indumentum, white at first turning to buff to rufous or greyish-white;calyx glandular ciliate .... *argipeplum*

4a. Leaves obovate to oval, upper surface conspicuously reticulate, lower surface with a sparse, thin, continuous, pale brown floccose and glandular indumentum; calyx glabrous .... *erosum*

R. argipeplum Balf.f. & Cooper, Notes R.B.G. Edinb. 9:213 (1916). Cox & Cox, The Encyclopedia of Rhododendron Species 25 (1997).

Syn.: *R. smithii* Nuttall ex Hook.f., Bot. Mag. 85 : t. 5120 (1859) non Sweet (1830).

Shrub or small tree, 2 - 7.5 m high. Young shoots covered with bristles or glandular-bristles. Bark smooth with a plum-coloured bloom, peeling. Leaves 8 - 16 x 2.7 - 6 cm, oblong lanceolate to elliptic-lanceolate, apex acute to rounded, base rounded, to obscurely cordate, margins reflexed, upper surface rich green and glossy with impressed main veins, lower surface with a thin continuous layer of pale brown indumentum of dendroid hairs, turning white on ageing; petioles 1 - 2 cm long, densely covered with gland-tipped bristles. Inflorescence dense, 10 - 20-flowered; pedicels 10 - 15 mm, glandular setose. Calyx fleshy, reddish, 5 - 10 mm, lobes rounded and glandular-ciliate. Corolla fleshy, tubular-campanulate, scarlet-crimson, 3 - 4.5 cm long, with nectar pouches. Ovary densely rufous-tomentose and glandular-hairy; style glabrous. Capsule *ca* 15 x 4 mm, straight.

Distribution and habitat : India : Sikkim, Arunachal Pradesh. Bhutan, China (S. Xizang). In mixed forests with other rhododendrons and fir, in the altitudinal range of 2,400 - 3,700 m. **Map 16.**

R. barbatum Wall. ex G. Don, Gen. Hist. 3:844 (1834). C.B. Clarke in Hook.f., Fl. Brit. India 3:468 (1882). Chamberlain, Notes R.B.G. Edinb. 39(2) : 380(1982). Cox & Cox, The Encyclopedia of Rhododendron Species 26 (1997).

Syn. : *R. lancifolium* Hook.f., Rhododendrons of Sikkim Himalaya t. 4 (1849).

Tree up to 10 m high, often branching from the base. Young shoots with long stiff bristles, rarely glandular and / or glabrous. Bark smooth, peeling, reddish or deep-purple coloured; growth buds sticky, bud scales red. Leaves 10 - 20 x 2.5 - 7 cm, elliptic-lanceolate, slightly convex, apex acute to acuminate, base rounded to cuneate, upper surface deep green, glabrous, rugulose, lower surface usually glabrous or sometimes with sparse, scattered dendroid hairs and stipitate glands; petioles 1 - 2 cm, usually covered with bristles. Inflorescence 10 - 20-flowered, compact; pedicels 5 - 10 mm, glabrous. Calyx cup-like, 10 - 15 mm, glabrous, lobes rounded, with erose margins. Corolla thick and fleshy, tubular-campanulate, up to 5 cm long, bright scarlet, with darker nectar pouches. Ovary densely glandular, sometimes with a few dendroid hairs; style glabrous. Capsule 15 - 20 x 5 mm, curved.

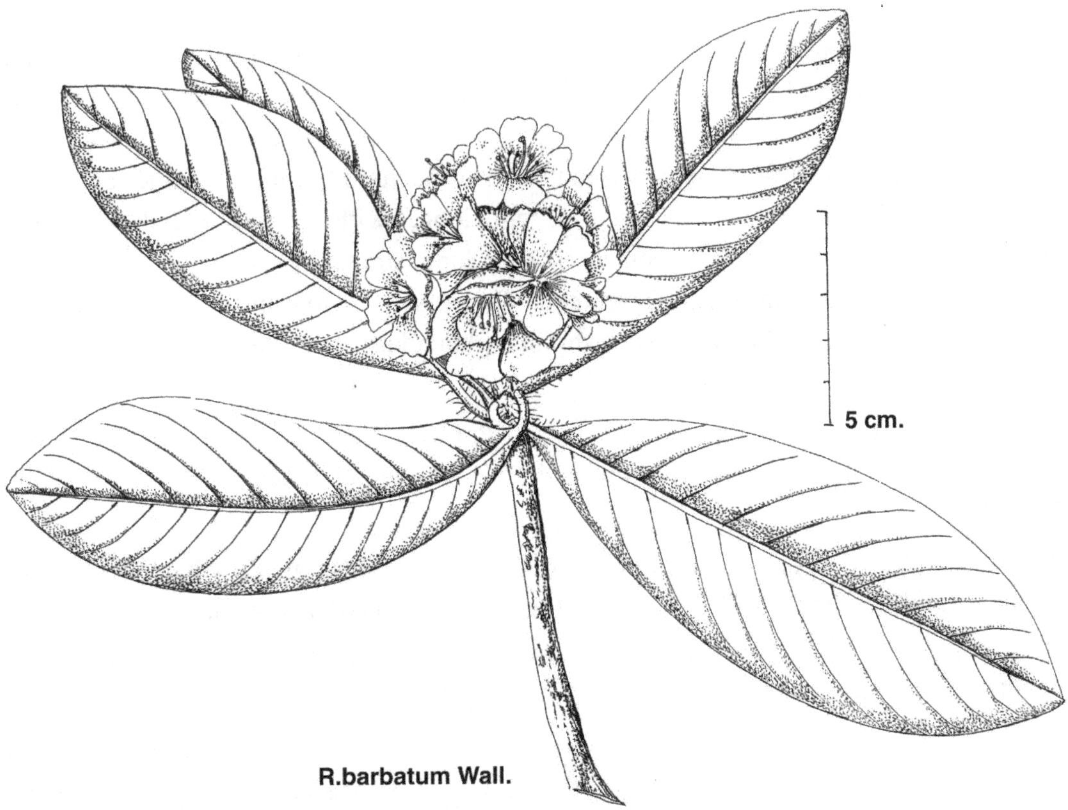

R.barbatum Wall.

Distribution and habitat: India : Himalayas in Uttaranchal, Darjeeling Hills, Sikkim, Western Arunachal Pradesh. Nepal, Bhutan, China (S. Xizang). Frequent in open slopes, mixed forests in the altitudinal range of 2,400 - 3,700 m. **Map 16.**

Notes: The species name *barbatum* refers to the 'bearded' or bristly young shoots, although this species exhibits much variation in the density and length of the bristles. In some western Himalayan forms, the bristly character is totally absent and were formerly considered as a separate species *R. imberbe* Hutchinson though now treated as synonymous to *R. barbatum* while Chamberlain considers this to be a hybrid between *R. barbatum* and *arboreum*. *R. barbatum* is a fine species with beautiful, smooth, reddish-plum coloured bark and deep red or bright-scarlet, dense flower trusses and is a gardeners' first choice. It is a parent of several merited clones that are in cultivation.

R. erosum Cowan, Notes R.B.G. Edinb. 19:225(1937). Chamberlain, Notes R.B.G. Edinb. 39(2) : 382(1982). Cox & Cox, The Encyclopedia of Rhododendron Species 26(1997).

Shrub or small tree, 3 - 9 m high. Young shoots with numerous gland-tiped setae. Bark brown, peeling. Leaves 8 - 20 x 3 - 10 cm, convex, obovate to oval, base cordulate, upper surface conspicuously reticulate, glabrous, lower surface with a sparse, thin continuous pale brown floccose and glandular indumentum, becoming thinner on ageing, midrib sparsely bristly; petioles *ca* 1 cm long, with a few glandular-bristles. Inflorescence 10 - 15-flowered; pedicels *ca* 1 cm long, stipitate-glandular. Calyx reddish, 3 - 5 mm, glabrous. Corolla tubular-campanulate, rose-pink to crimson, 3 - 3.5 cm long, with nectar pouches. Ovary densely stipitate-glandular; style glabrous. Capsule 2 cm long, curved.

Distribution and habitat: India : Arunachal Pradesh. E. Bhutan to China (S. & S.E. Xizang). In *Rhododendron-Abies* forests in the altitudinal range of 3000 - 4000 m. **Map 16.**

Notes: An attractive species with Copper to plum coloured young growth. The species is included on the authority of Mr. P.A. Cox (pers. comm. to ARK Sastry).

R. exasperatum Tagg, Notes R.B.G.Edinb. 16:192 (1931). Chamberlain, Notes R.B.G. Edinb. 39(2) : 382(1982). Cox & Cox, The Encyclopedia of Rhododendron Species 27 (1997).

Shrub or small tree, 2 - 5 m high. Young shoots bronze and densely covered with glandular bristles and persistent shaggy perulae. Bark peeling, reddish-brown. Leaves

9 - 17 x 4.5 - 10 cm, broadly obovate to elliptic, convex, rounded at both ends, upper surface dark green, glabrous and shiny, lower surface with sparse stout gland-tipped bristles, bristles larger along the midrib; petioles 5 - 10 mm, long bristly. Inflorescence 10 - 15-flowered, dense; pedicels *ca* 15 mm long, sparsely stipitate-glandular. Calyx 3 - 5 mm, reddish, lobes rounded. Corolla tubular-campanulate, 3.5 - 4.5 cm long, brick-red, with nectar pouches. Ovary densely stipitate-glandular; style glabrous.

Distribution and habitat : India : Delei valley in Eastern Arunachal Pradesh. N.E. Burma, China (S.E. Xizang). Rather rare, in *Abies* forests, on rocks in thickets, and on open steep slopes, in the altitudinal range 2,900 - 3,700 m. **Map 16.**

Notes: *R. exasperatum* meaning 'rough' on account of its dense, stiff, gland-tipped bristles on young shoots, is a distinctive species with large, broad, stiff leaves with glandular bristles on the lower surface and bronze coloured new growth. The species is grown both for its attractive foliage and brick-red flowers.

R. succothii Davidian, Rhododendron Yearbook 20:103, t.8 (1965). Chamberlain, Notes R.B.G. Edinb. 39(2) : 382(1982). Cox & Cox, The Encyclopedia of Rhododendron Species 28(1997).

Shrub or small tree, 1 - 6 m high. Young shoots glabrous. Bark red-brown or purplish grey, peeling. Leaves 5 - 13 x 2.5 - 6.5 cm, oblong to elliptic, apex rounded and minutely apiculate, base cordate, whorled at twig-ends, upper surface dark green, glossy, glabrous, lower surface mostly glabrous, rarely hairy; petioles very short, glabrous. Inflorescence 10 - 15-flowered, dense; pedicels 5 - 15 mm long, glabrous. Calyx minute, gabrous. Corolla tubular-campanulate, 2.5 - 3.5 cm long, crimson or scarlet, with pronounced nectar pouches at the base. Ovary and style glabrous. Capsule 20 x 5 mm, curved.

Distribution and habitat : India : Kameng and Tawang districts in W. Arunachal Pradesh. Bhutan. Rare, in mixed forests amongst other shrubby rhododendrons and along forest margins, in the altitudinal range of 3,400 - 4,100 m. **Map 16.**

Notes: The only species in the subsection *Barbata* totally devoid of the characteristic bristles on the young growth and the indumentum on the leaf undersurface.

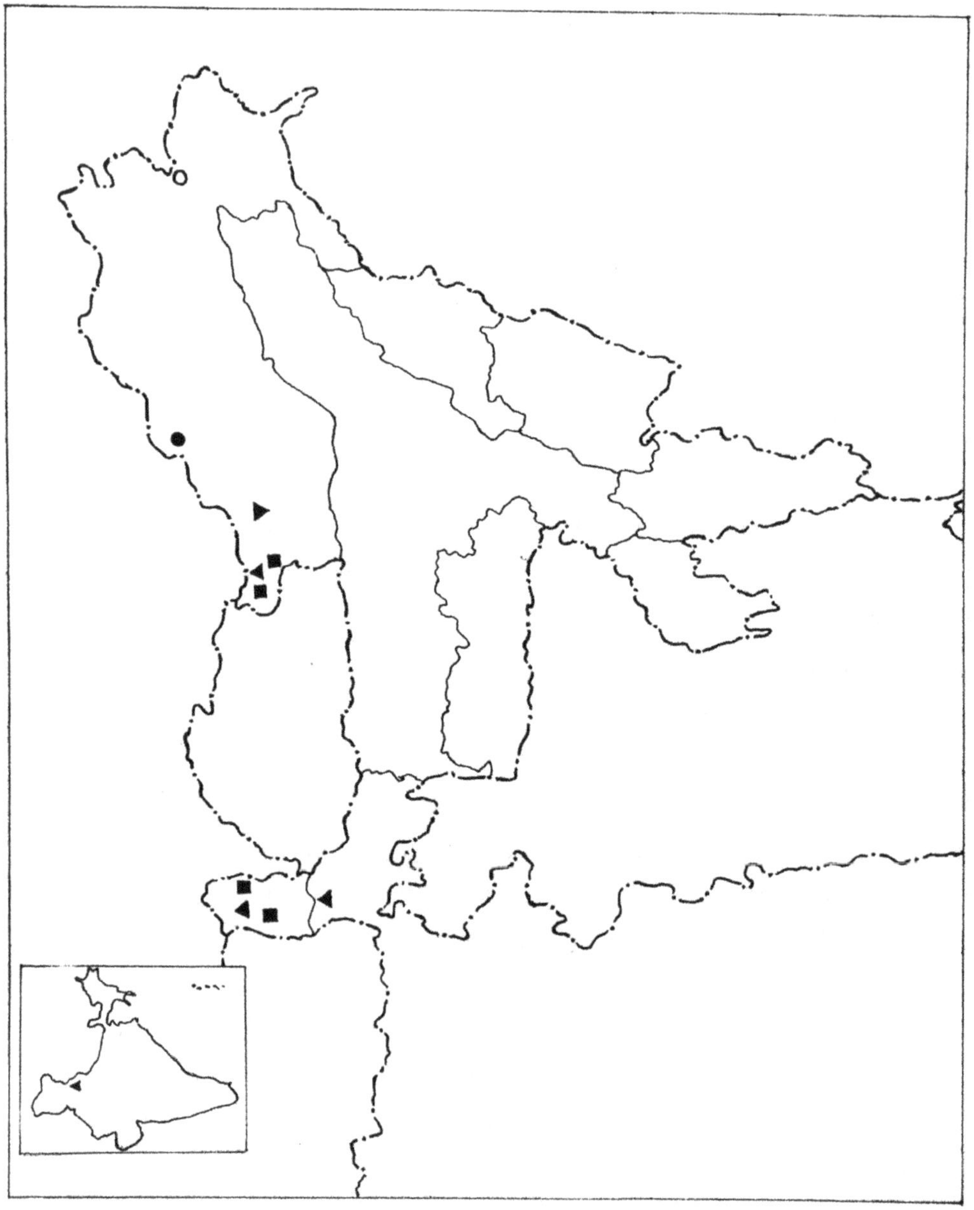

Map 16. ■ R.agripeplum. ▲ R.barbatum. ● R.erosum. O R.exasperatum. ▼ R.succothii.

(Maps are only indicative and not to scale)

Subsection Campanulata

Subsection **Campanulata** Sleumer, Bot. Jahrb. 74:550 (1949). Chamberlain, Notes R.B.G. Edinb. 39(2) : 371(1982). Cox & Cox, The Encyclopedia of Rhododendron Species 29 (1997).

Syn.: Series *Campanulatum* sensu Tagg in Stevenson (ed.) The Species of Rhododendron 175 (1930) pro parte.

Shrubs or small trees, 0.5 - 11 m high. Young shoots glabrous or with whitish floccose tomentum. Bark grey, fawn to pinkish-brown, smooth or rough on ageing. Leaves 5 - 18 x 2 - 9 cm, ovate to broadly elliptic, upper surface shiny to glaucous, lower surface with a dense, fawn rusty brown indumentum, or a loose or scattered indumentum or nearly glabrous at times. Inflorescence 5 - 18-flowered; rachis 1 - 2.5 cm long. Calyx 1 - 3 mm, sparsely tomentose or glabrous. Corolla open to funnel-campanulate, 5-lobed, white, cream, pink, pinkish-red, rosy-purple, blue-lavender, spotted or not, without nectar pouches. Stamens 10. Ovary and style almost glabrous.

Type species: *R. campanulatum* D. Don

Key to the species of subsection Campanulata

1. Indumentum on the lower leaf surface densely fulvous, continuous, composed of branched hairs .... *campanulatum*

1a. Indumentum on the lower leaf surface sparse, dark brown, discontinuous, composed of fasciculate hairs .... *wallichii*

R. campanulatum D. Don, Mem. Wern. Nat. Hist. Soc. 3:410 (1821). C.B. Clarke in Hook.f., Fl. Brit. India 3:466 (1882). Chamberlain, Notes R.B.G. Edinb. 39(2) : 372 (1982). Cox & Cox, The Encyclopedia of Rhododendron Species 29(1997).

Shrub or small tree. Branchlets usually glabrous. Bark grey to fawn, rough on ageing. Leaves 7 - 18 x 2.5 - 9 cm, ovate to broadly elliptic, apex rounded, apiculate, base rounded to cordate, upper surface glabrous when mature, glossy, lower surface with a continuous dense, fawn to rusty brown lanate tomentum composed of capitellate to ramiform (branched) hairs; petioles 1.5 - 2.5 cm, glabrous. Inflorescence 6 - 18-flowered; pedicels 1.5 - 2.5 cm, glabrous. Calyx up to 2 mm, glabrous. Corolla broadly campanulate, 3 - 4.5 cm long, various shades of purple, rosy-white or white, spotted with purple. Ovary and style glabrous. Capsule 2 - 3 cm long, usually curved.

The species name *companulatum* refers to its bell-shaped flowers.

Key to the subspecies of R. campanulatum

1. Leaves 9.5 - 18 cm long, subcoriaceous, upper surface without a glaucous bloom, lower surface with a continuous, unistrate indumentum; corolla white to pale rose or lilac .... subsp. *campanulatum*

1a. Leaves 7 - 10 cm long, coriaceous, upper surface with a glaucous metallic bloom at first, lower surface with a buff to rusty-brown, bistrate indumentum; corolla pink to purple.. subsp. *aeruginosum*

R.campanulatum D.Don

R. campanulatum D. Don subsp. **aeruginosum** (Hook.f.) Chamberlain, Notes R.B.G. Edinb. 37:329 (1979) & Notes R.B.G. Edinb. 39(2) : 373(1982). Cox & Cox, The Encyclopedia of Rhododendron Species 29(1997). **(Photo No. 51).**

Syn.: *R. aeruginosum* Hook.f., Rhododendrons of Sikkim Himalaya 23, t. 22 (1849).
R. campanulatum D. Don var. *aeruginosum* Hook.f. ex Cowan & Davidian, Rhododendron Yearbook 4:168(1949).

Small shrub, usually up to 2 m (rarely more) high. Leaves smaller than in subsp. *campanulatum,* 7 - 10 cm long, upper surface with a glaucous metallic bloom in the young stages whch may partially persist in older leaves, lower surface with a rusty-brown, bistrate, thick tomentum. Corolla pink to purple.

Distribution and habitat : India : Sikkim. Bhutan. Less frequent and in alpine areas in the altitudinal range of 3,700 - 4,500 m. **Map 17.**

Notes: This subspecies with its attractive foliage and pinkish purple flowers is a gardener's choice.

R. campanulatum D. Don subsp. **campanulatum. (Photo Nos. 48, 49).**

Shrub or small tree, 2 - 11 m high. Leaves 9.5 - 18 cm long, subcoriaceous, upper surface glabrous on maturity and without a glaucous bloom, lower surface with a continuous unistrate indumentum. Corolla white **(Photo No. 50)** to pale rose or lilac.

Distribution and habitat : India : Kashmir to Sikkim. Nepal, Bhutan. Frequent in mixed forests and scrub jungles in the altitudes of 3,000-4,500 m. **Map 17.**

R. wallichii Hook.f. Rhododendrons of Sikkim Himalaya t.5 (1849). Chamberlain, Notes R.B.G. Edinb. 39(2) : 373(1982). Cox & Cox, The Encyclopedia of Rhododendron Species 31 (1997). **(Photo No. 52).**

Syn.: *R. campanulatum* Hook.f. var. *wallichii* (Hook.f.) C.B. Clarke in Hook.f., Fl. Brit. India 3:466 (1882).

Bushy shrub or a small tree, 2 - 6 m high. Young shoots with a floccose tomentum. Bark pinkish-brown to grey, rough on ageing. Leaves leathery, 7 - 11 (- 14) x 3.5 - 5.5 cm, elliptic to ovate, apiculate and rounded at apex, rounded or subcordate at base, upper surface dark and shiny, glabrous or with a patchy layer of black to brown, tufted hairy indumentum ; petioles 1 - 2 cm long, covered with a floccose tomentum. Inflorescence 5 - 10-flowereed, rather lax; pedicels 1 - 1.5 cm, nearly glabrous or sparsely covered by tufted hairs. Calyx irregular, sparsely tomentose or glabrous, lobes 1 - 3 mm long. Corolla funnel-campanulate, 4 - 5 cm long, blue-mauve or pale mauve or rosy-purple fading to white, spotted or not. Ovary glabrous; style glabrous. Capsule 15 - 30 x 5 - 7 mm.

Distribution and habitat : India : Sikkim, W. Arunachal Pradesh. E. Nepal, Bhutan, China (S. Xizang). In mixed forests and Rhododendron scrub in sub-alpine to alpine areas above tree-line in the altitude of 3,000 - 4,300 m. **Map 17.**

Notes: The species is named after the botanist Nathaniel Wallich, a one time Superintendent of the Botanic Garden at Calcutta and famous for his Catalogue of Plants popularly called *Wallich Catalogue.*

Subsection Campylocarpa

Subsection **Campylocarpa** Sleumer, Bot. Jahrb. 74:547 (1949). Chamberlain, Notes R.B.G. Edinb. 39(2) : 262(1982). Cox & Cox, The Encyclopedia of Rhododendron Species 32 (1997).

Syn.: Series *Thomsonii* subseries *Campylocarpum* and subseries *Souliei* sensu Tagg in Stevenson (ed.) The Species of Rhododendron, 698 & 725 (1930).

Only the Type species *R. campylocarpum* Hook.f. subsp. *campylocarpum* occurs in India.

R. campylocarpum Hook.f., Rhododendrons of Sikkim Himalaya, t. 30 (1851) subsp. **campylocarpum.** Chamberlain, Notes R.B.G. Edinb. 39(2) : 263(1982). Cox & Cox, The Encyclopedia of Rhododendron Species 33 (1997). **(Photo Nos. 53-55).**

Shrub or small tree, to 6 m high. Young shoots usually with a few short stipitate glands. Leaves 1.6 - 2.5 x as long as broad, elliptic, rounded and shortly apiculate at apex, more or less cordate at base, upper surface glabrous, dark glossy green, lower surface pale glaucous-green, nearly glabrous; petioles up to 2.2 cm long, stipitate-glandular when young. Inflorescence 3 - 10 (- 15)-flowered; pedicels 10 - 35 mm, stipitate-glandular. Corolla campanulate, 2.5 - 4 cm long, pale to bright yellow, with or without a basal blotch. Ovary densely-stipitate glandular; style glabrous or glandular towards the base, eglandular in the upper half. Capsule 1.3 - 2 cm long, slender, curved.

Distribution and habitat : India : Sikkim & Arunachal Pradesh. E. Nepal, Bhutan, China (S.E. Xizang). Frequent in open forests with other rhododendrons and among bamboo thickets, in the altitudinal range of 2,900 - 4,300 m. **Map 17.**

Notes: *campylocarpum* meaning with bent fruits, is certainly one of the finest yellow-flowered species and is a parent of many prized clones commonly grown in the gardens in the West.

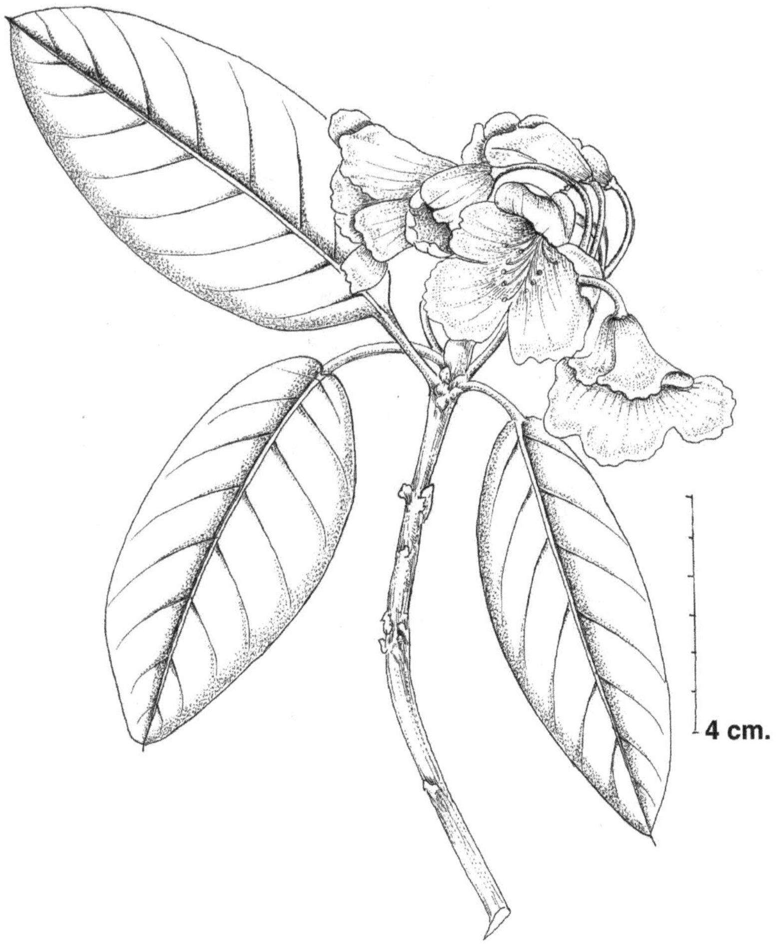

R.campylocarpum Hook.f.

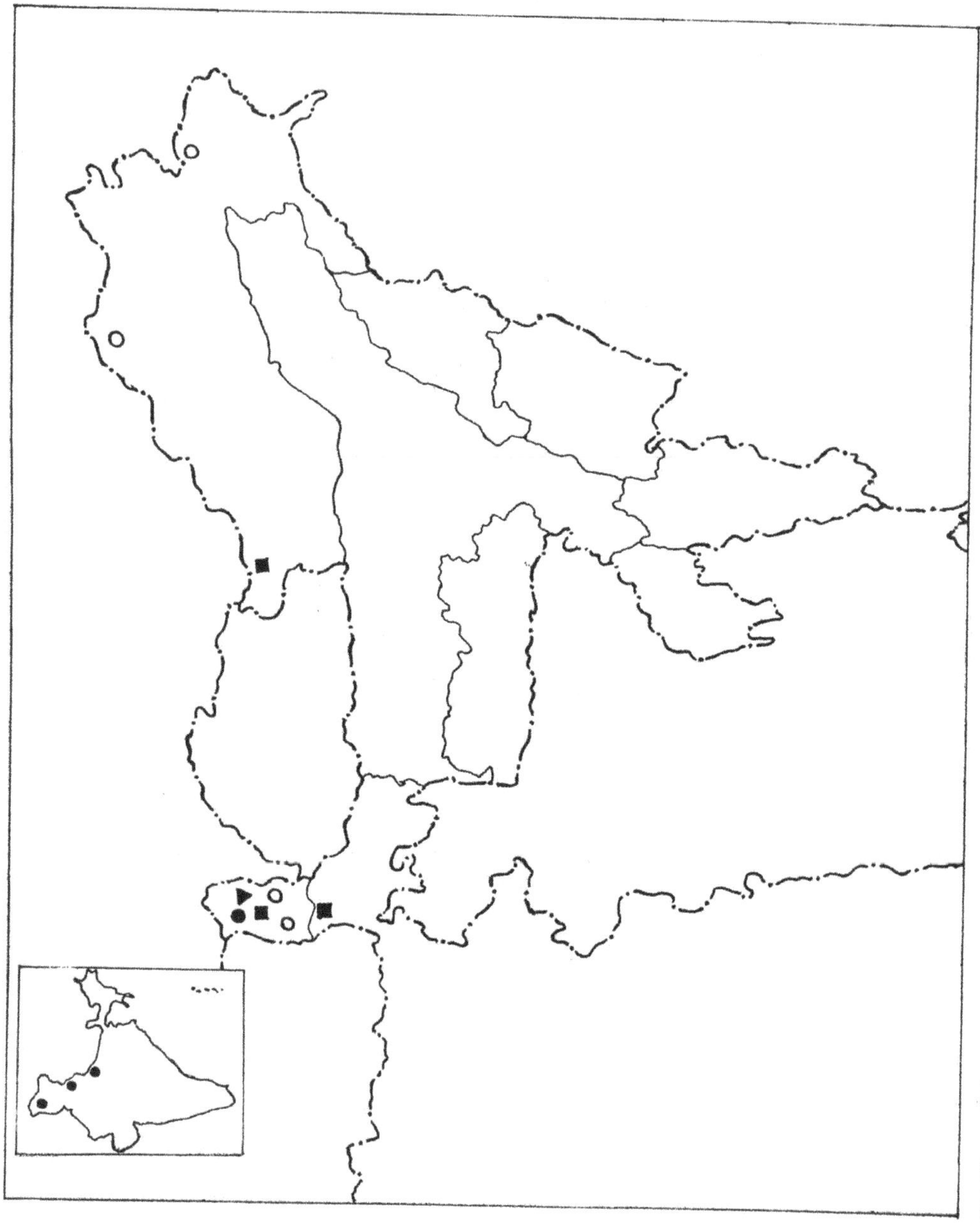

Map 17. ● R.campanulatum ssp.campanulatum. ▼ ssp.aeruginosum. ■ R.wallichii. O R.campylocarpum ssp.campylocarpum.

(Maps are only indicative and not to scale)

Subsection Falconera

Subsection **Falconera** Sleumer, Bot. Jahrb. 74:548 (1949). Chamberlain, Notes R.B.G. Edinb. 39(2) : 251(1982). Cox & Cox, The Encyclopedia of Rhododendron Species 37(1997).

Syn.: Series *Falconeri* sensu Tagg in Stevenson (ed.) The Species of Rhododendron 235 (1930).

Large shrubs or trees with stout trunks, 3 - 14 m high. Bark rough or more or less smooth, grey, pink to reddish-brown, flaking or peeling. Banchlets usually tomentose. Leaves up to 46 cm long, 17 cm broad, oblanceolate to broadly ovate, thick, leathery, upper surface glabrous or with semi-persistent indumentum, lower surface with dense whitish to rufous woolly indumentum often with cup shaped hairs. Inflorescence 10-30-flowered; rachis (5 -) 10 - 60 mm. Calyx minute, 1 - 3 mm. Corolla (5 -) 7 - 10-lobed, funnel-to oblique or ventricose-campanulate, nectar pouches absent, yellow or white to rose-purple, often blotched or spotted. Stamens 12 - 18. Ovary tomentose, glandular or glabrous; style glabrous.

Type species: *R. falconeri* Hook.f.

Notes: Subsection *Falconera* species are closely related to the species of Subsection *Grandia* but differ in the absence of nectar pouches and in having woolly indumentum comprising of distinctive cup-shaped hairs as against the plastered indumentum in most of the species in subsection *Grandia.*

Subsection *Falconera* includes some of the most elegant species of eminence with attractive foliage and matching floral trusses and are most popularly grown in U.K and other temperate gardens.

Key to the species of subsection Falconera

1. Ovary and pedicels densely viscid-glandular, not-tomentose .... *falconeri*

1a. Ovary and pedicels not glandular, densely tomentose 2

2. Corolla bright, dark pink or rose-purple, without flecks; leaf indumentum silvery to cinnamon .... *hodgsonii*

2a. Corolla white or pale pink, usually with flecks; leaf indumentum fawn to rufous or rich rufous-brown .... *arizelum*

R. arizelum Balf.f. & Forrest, Notes R.B.G. Edinb. 12:90 (1930). Cox & Cox, The Encyclopedia of Rhododendron Species 37(1997). **(Photo No. 56).**

Syn.: *R. rex* Leveille', Feddes Repert. 13:340 (1914) subsp. *arizelum* Chamberlain, Notes R.B.G. Edinb. 36:330 (1979).

A flat-topped tree upto 7 m high. Young branches covered with cinnamon or greyish tomentum. Bark pink to reddish-brown. Leaf bud scales cone-shaped. Leaves 7.5 - 25 x 3.3 - 12 cm, obovate to oblanceolate, nearly rounded at apex, cuneate at base; upper surface dark green and rugulose with semi-persistent indumentum, lower surface with thick, woolly, persistent, cinnamon or brown indumentum comprising strongly fimbriate, cup-shaped hairs; petioles 2 - 3 cm long, terete, with a dense pale tomentum. Inflorescence 15 - 20-flowered, compact; pedicels up to 3 cm long, brownish-tomentose. Calyx *ca* 2 mm, densely tomentose. Corolla fleshy, obliquely campanulate, *ca* 4.5 cm long, 8-lobed, cream, yellow, rose, mauve, crimson or carmine coloured. Stamens 15 - 16. Ovary densely brown-tomentose.

Distribution and habitat: India : Subansiri District, Arunachal Pradesh. China (S.E. Xizang & N.W. Yunnan). In Rhododendron forests on a ridge at 3000 m. **Map 18.**

Notes: This species closely resembles *R. falconeri* subsp. *eximium* but differs in its more obovate, less matt-rugulose leaves, the eglandular pedicels and ovary, and usually in its different flower colour. Both *R. arizelum* and *R. falconeri* subsp. *eximium* have been collected for the first time in India in April 1965 in dense Rhododendron forest growing on a ridge at 3,000 m altitude, in the south east direction of Hapoli, Apathani valley in Subansiri district during one of the authors' (A.R.K.S) joint Botanical Exploration with Mr. P.A. Cox, Sir Peter Hutchison and Mrs. Patricia Cox and reported as new distribution record for India.

The specific epithet *arizelum* means notable. This is a fine foliage plant and the forms with deep coloured flowers are considered best in horticulture.

R. falconeri Hook.f., Rhododendrons of Sikkim Himalaya t. 10 (1849). Chamberlain, Notes R.B.G. Edinb. 39(2) : 259(1982). Cox & Cox, The Encyclopedia of Rhododendron Species 40(1997).

Large shurb or tree, 6 - 15 m tall with reddish-brown, rough flaking bark. Young shoots covered with grey-fawn tomentum. Leaves thick, leathery, 15 - 35 x 7 - 17 cm, broadly elliptic to broadly obovate, apex rounded, base rounded to cordate, upper surface dark matt green, rugulose, glabrous or with scurfy indumentum, nerves deeply impressed, lower surface with a dense, bistrate, rust-coloured indumentum, the lower layer somewhat white, the upper layer with strongly fimbriate, narrowly cup-shaped hairs; petiole terete, 2.5 - 5 cm, sparsely floccose and stipitate-glandular. Inflorescence 15 - 20 (or more)-flowered, compact; rachis thick, 30 - 60 mm; pedicels 40 - 55 mm, densely stipitate-glandular, viscid. Calyx *ca* 2 mm, viscid-glandular. Corolla thick, 8(- 10)-lobed, obliquely-campanulate, 4 - 6 cm long, creamy-white to pale yellow or pale pink with a purple blotch at base. Stamens 10 - 16. Ovary densely viscid-glandular; style thick, glabrous or tomentose, glandular at base. Capsule *ca* 40 x 10 mm, straight.

Notes: This species differs from its nearest relative *R. arizelum* in its larger size of the trees, larger leaves with heavily matted upper surfaces and in its glandular ovary.

R.falconeri Hook.f.

Two subspecies of R. falconeri are recognized as follows:

1. Corolla white to cream ; leaves glabrous at maturity .... subsp. *falconeri*

1a. Corolla pale pink with darker tips; leaves scurfy above even at maturity .... subsp. *eximium*

R. falconeri Hook.f. subsp. **falconeri** Chamberlain, Notes R.B.G. Edinb. 39(2) : 260(1982). Cox & Cox, The Encyclopedia of Rhododendron Species 40 (1997). **(Photo Nos. 57-60).**

Taller and more robust than subsp. *eximium* with upper surface of leaves becoming glabrous on ageing and flowers with white to cream to pale yellow colour having a purple blotch at the base and the stamens ranging from 12 - 16 in number.

Distribution and habitat: India : Sikkim, Darjeeling Hills, West Bengal, Western Arunachal Pradesh. E. Nepal & Bhutan. Frequent in mixed wet temperature forests and amongst conifers, at altitudes ranging from 2,700 - 3,400 m. **Map 18.**

Notes: Named after Hugh Falconer, a former Superintendent of Sahranpur Botanic Gardens in 1832, this species is one of the most gorgeous amongst rhododendron species and is a parent of prize winning clones grown in the British Gardens.

R. falconeri Hook.f. subsp. **eximium** (Nuttall) Chamberlain, Notes R.B.G. Edinb. 37:330 (1970) & Notes R.B.G. Edinb. 39(2) : 260(1982). Cox & Cox, The Encyclopedia of Rhododendron Species 40(1997).

Syn.: *R. eximium* Nuttall, Hooker's J. Bot. Kew Gard. Misc. 5:364 (1853).

A large shrub, or smaller tree (than subsp. *falconeri*). Young shoots and foliage buds covered with rusty tomentum. Leaves oval to obovate-elliptic, dark green with deeply impressed nerves and semi-persistent rusty-brown indumentum on the upper surface and with deeper cinnamon-coloured indumentum on the lower surface. Corolla fleshy, rose to cream flushed rose at first, fading out to creamy-pink. Stamens 10 - 14.

Distribution and habitat: India : Arunachal Pradesh. Bhutan. Rather rare; amidst other rhododendrons in mixed, temperate wet evergreen forests and on mountain ridges in the altitudinal range of 2700 - 3400 m. **Map 18.**

Notes: This subspecies is very close and related to and easily confused with *R. arizelum* which has generally smaller, less oval and more obovate leaves, less persistent indumentum on the upper leaf surface, eglandular ovary and flower colour varying from cream, pale yellow to rose. In the subspecies *eximium* the ovary is glandular and the flowers are rose fading to cream, besides the larger leaves with semi-persistent indumentum on the upper leaf surface. *Eximium* means excellent; is a fine foliage plant.

One of the authors (A.R.K.S) collected both *R. falconeri* subsp. *eximium* and *R. arizelum* growing together in mixed temperate broad-leaved forests along the ridge at 2,700 m altitude, ca 40 km S.E. of Hapoli in Subansiri district, Arunachal Pradesh during a joint botanical exploration with Mr. P.A. Cox and Sir P. Hutchison in April 1965.

R. hodgsonii Hook.f. Rhododendrons of Sikkim Himalaya 16, t. 15 (1851). C.B.Clarke in Hook.f., Fl. Brit. India 3:464 (1882). Chamberlain, Notes R.B.G. Edinb. 39(2) : 259(1982). Cox & Cox, The Encyclopedia of Rhododendron Species 43 (1997). **(Photo Nos. 61, 62).**

Shrub or often a small tree up to 10 m. Bark smooth, creamy to reddish-brown, peeling. Branchlets thick. Leaf buds conical with long tailed bud scales. Leaves 18 - 38 x 7 - 14 cm, ovate to oblanceolate or oblong-elliptic, leathery, apex rounded to retuse, base cuneate-rounded, upper surface dark green with a thin indumentum, creating a metallic bloom, lower surface with a bistrate (two-layered), dense, silvery to cinnamon indumentum, the lower (inner) layer compacted, the upper (outer) layer composed of broadly cup-shaped hairs; petioles 2.5 - 5 mm, terete, with thin floccose, greyish indumentum. Inflorescence compact, 15 - 25-flowered ; pedicels 2 - 4 cm long, covered with a sparse greyish floccose indumentum. Calyx 2 - 3 mm, tomentose. Corolla fleshy, tubular-campanulate, 3 - 4 cm long, 7 - 8-lobed, pink to magenta or purple, usually with a darker basal blotch. Stamens 15 - 18. Ovary tomentose. Capsule 3 - 4 cm long, *ca* 0.6 mm thick, curved, sub-tomentose.

Distribution and habitat: India : Sikkim, Darjeeling in West Bengal, Arunachal Pradesh. E. Nepal, Bhutan, China (S. Xizang). Frequent in mixed temperate forests on ridges, forested valleys and in conifer forests often associated with *Abies pindrow* at altitudes of 3,000 - 4,000 m. **Map 18.**

Notes: This species is readily distinguislhed by its smooth, creamy or often cinnamon coloured peeling brak; characteristically cone-shaped buds with apices drawn into long tails; metallic looking upper leaf surface; distinctive silvery to cinnamon indumentum on the leaf lower surface and compact trusses of deep purplish flowers. A much sought after species in horticulture especially for its fine, attractive smooth bark and foliage and very pleasing flower trusses.

Wherever this species is found growing in the wild in association with *R. falconeri,* stable populations of natural hybrids between the two are of frequent occurrence. *R. x decipiens* Lacaita *(Journ. Linn. Soc. Bot.* 43:473, 1916) described from specimens collected from North Bengal and Sikkim, is considered to be one such natural hybrid (Chamberlain, *Notes R.B.G. Edinb.* 39(2) : 259(1982). Cox & Cox, The *Encyclopedia of Rhododendron Species* 43 (1997). **(Photo Nos. 63, 64).**

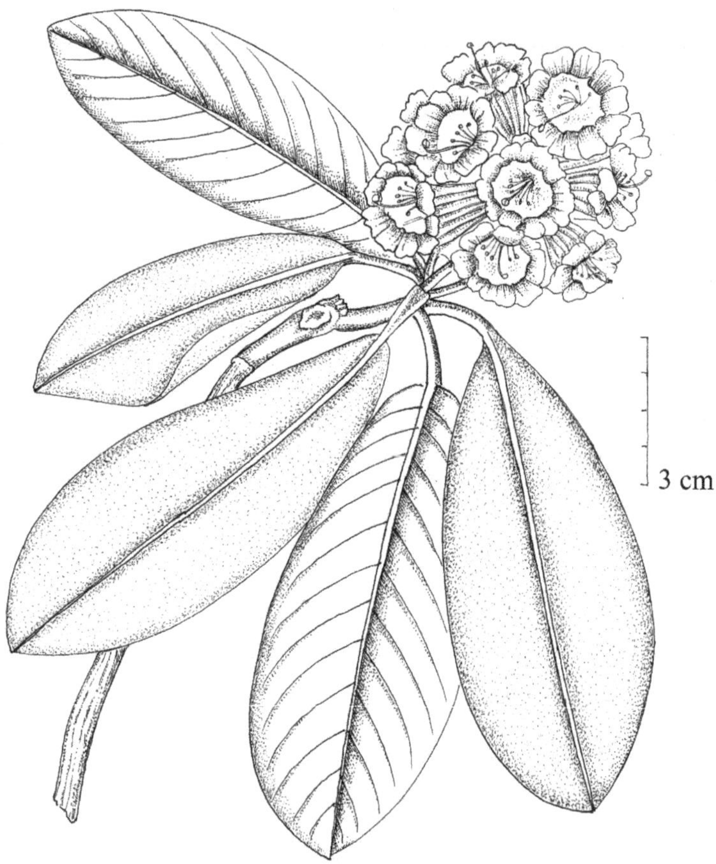

R.hodgsonii Hook.f.

The species is named after B.H. Hodgson, a one time Resident Officer in Nepal of the former East India Company.

Subsection Fortunea

Subsection **Fortunea** Sleumer, Bot. Jahrb. 74:546 (1949). Chamberlain, Notes R.B.G. Edinb. 39(2) : 224 (1982). Cox & Cox, The Encyclopedia of Rhododendrons Species 49 (1997).

Syn.: Series *Fortunei* sensu Tagg in Stevenson (ed.) The Species of Rhododendron 257 (1930).

Shrubs or small trees, 1.5 - 8.0 m. Bark rough, brown to grey. Young shoots at first covered with a thin, whitish to grey floccose indumentum, soon glabrescent. Leaves oblanceolate, elliptic or ovate, upper surface glabrous, lower surface with a persistent floccose indumentum on the midrib. Inflorescence 5 – 30-flowered, lax. Calyx usually small or well-developed (*R. griffithianum*). Corolla funnel-campanulate to open campanulate, often glandular 5 - 7 (- 8)-lobed, white to pink or purplish-pink, often scented. Stamens 12 - 25. Ovary stipitate-glandular or glabrous; style glabrous or stipitate-glandular to tip.

Type species: *R. fortunei* Lindley

This subsection is so far represented in India only by *R. griffithianum.*

R. griffithianum Wight, Ic. Pl. Ind. Or. 4:6, t. 1203 (1850). C.B.Clarke in Hook.f., Fl. Brit. India 3:468 (1882). Chamberlain, Notes R.B.G. Edinb. 39(2) : 237(1982). Cox & Cox, The Encyclopedia of Rhododendron Species 55 (1997). **(Photo Nos. 65, 66).**

Syn.: *R. aucklandii* Hook.f., Rhododendrons of Sikkim Himalaya t. 11 (1851). *R. oblongum* Griffith, Ic. Pl. Asiat. t. 523 (1854).

Small tree, 12 - 15 m tall in the wild. Bark smooth, thin, light reddish-brown, flaking and peeling, young shoots pale green. Leaves 10 - 30 x 4 - 10 cm, oblong, oblong-elliptic or oblong-ovate, surfaces glabrous, apex acute to rounded, shortly mucronate, base rounded, yellowish-green with a glaucous tint on the lower surface; petiole 2.5 - 5.0 cm long, glabrous. Inflorscence 3 - 6-flowered; rachis up to 8 cm long, thick, slightly glandular; pedicels 3.5 - 4.0 cm long, sparsely glandular. Calyx saucer-shaped, large, 1 - 2 cm, green or pink tinged, irregular, enlarged in fruit. Corolla 5-lobed, very large, widely campanulate, glabrous, 3 - 7 cm long and up to 15 cm wide, delicately fragrant, white, sometimes with green spotting or white tinged with pink, beautifully veined red.

Stamens 12 - 18, filaments glabrous. Ovary glandular; style glandular-pubescent. Capsule 2 - 4 cm x 1.3 - 1.8 cm, stout, short-cylindric, woody, at length glabrous. Seeds ellipsoid, compressed.

Distribution and habitat: India : Sikkim, Darjeeling Hills in West Bengal, Arunachal Pradesh. Nepal, Bhutan. Frequent in moist temperate forests mixed with other tree-rhododendrons, magnolias and oaks, in the altitudinal range of 1800 - 2900 m. **Map 18.**

Notes: A magnificent species with beautiful smooth often reddish-brown bark, very attractive large flowers on long rachis and well-developed saucer-shaped calyx.

The species has amongst the largest flowers in the genus and has been used a very great deal in hybridization and is a parent of several award winning clones.

The species is named after William Griffith, a botanist-explorer and former Superintendent of the Botanical Garden at Sibpur, Calcutta.

Subsection Fulgensia

Subsection **Fulgensia** Chamberlain, Notes R.B.G. Edinb. 37:336(1979).

Shrubs or small trees, 1.5 - 6 m high. Bark smooth, peeling. Leaves thick, leathery, elliptic or broadly ovate, lower surface with a dense, fulvus, lanate-tomentum consisting of fasciculate hairs. Inflorescence 8 - 14-flowered. Calyx minute, sometimes well-developed and cup-like. Corolla 5-lobed, funnel to tubular-campanulate, with nectar pouches, crimson to deep carmine. Stamens 10. Ovary and style glabrous.

Type species: *R. fulgens* Hook.f.

Some authors consider this to be a monotypic subsection.

R. fulgens Hook.f., Rhododendrons of Sikkim Himalaya 27, t.25(1849). Chamberlain, Notes R.B.G. Edinb. 39(2) : 415(1982). Cox & Cox, The Encyclopedia of Rhododendron Species 63(1997). **(Photos No. 67).**

Shrub or small tree. Branchlets glabrous, with bright crimson-coloured bracts. Bark pinkish-grey to reddish-brown, smooth, peeling. Leaves 5 - 13 x 2.5 - 7 cm, broadly ovate or elliptic, apex rounded, base rounded to cordate, young leaves glaucous above becoming dark green and shiny at maturity, lower surface with a thick fawn to fulvous, unistrate indumentum. Petiole up to 2 cm long. Inflorescence 8 - 14-flowered, compact and rounded; pedicels up to 1 cm long, glabrous. Calyx small, 1 - 2 mm, glabrous. Corolla tubular-campanulate, scarlet to deep red, with dark nectar pouches, 2 - 3.3 cm long. Stamens 10. Ovary and style glabrous. Capsule curved.

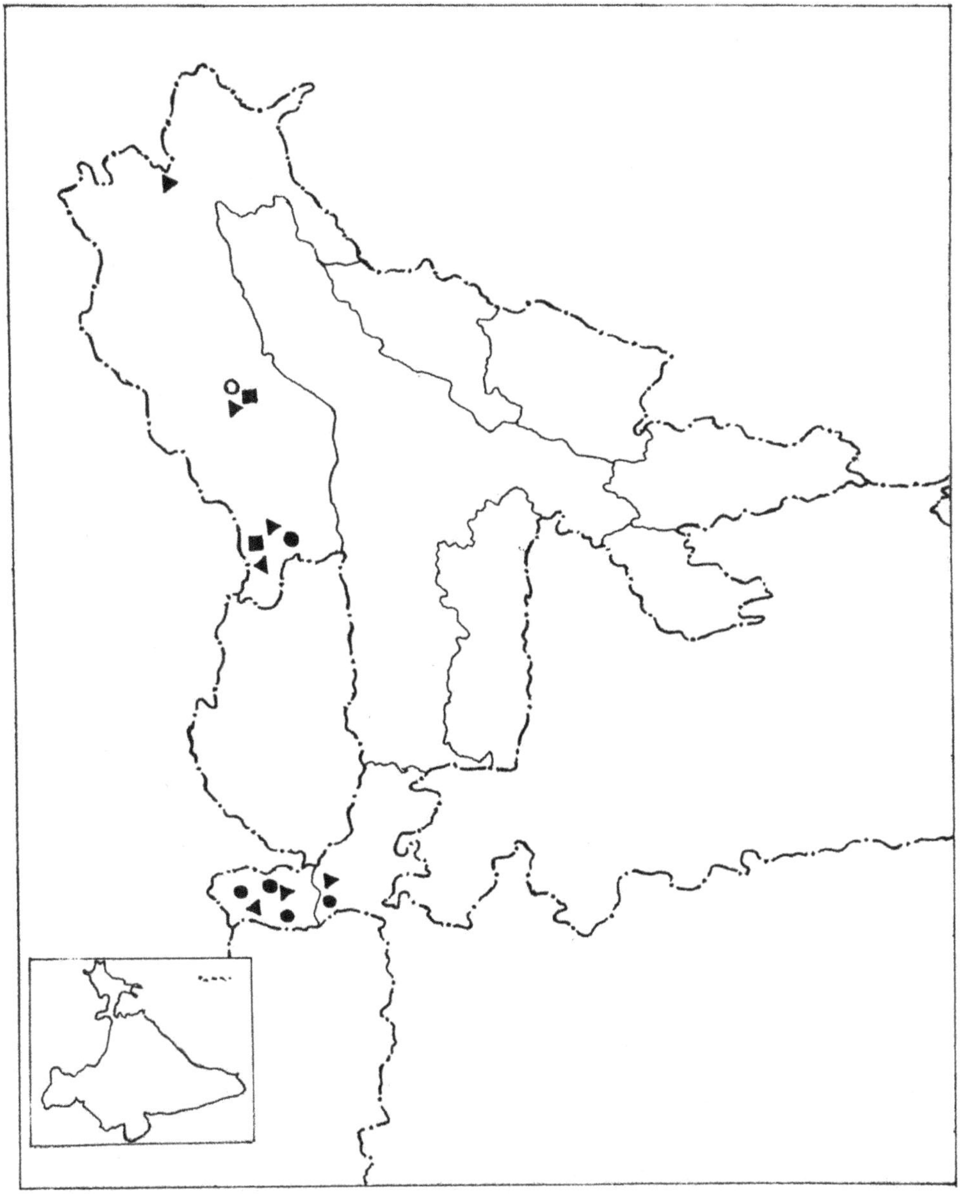

Map 18. ○ R.arizelum. ● R.falconeri ssp.falconeri. ■ ssp.eximium. ▲ R.hodgsonii. ▼ R.griffithianum.

(Maps are only indicative and not to scale)

Distribution and habitat : India : Sikkim, Darjeeling hills, W. Arunachal Pradesh. E. Nepal, Bhutan, China (S. Xizang). In mixed forests, in the altitudes of 3,200 - 4,300 m. Rather rare. **Map 19.**

Notes: The species with fine foliage, brilliant crimson bracts and compact flower trusses presents a shiny appearance *(fulgens* = shining) and is a gardener's favourite.

R.fulgens Hook.f.

Subsection Fulva

Subsection **Fulva** Sleumer, Bot. Jahrb. 74:549 (1949). Chamberlain, Notes R.B.G. Edinb. 39(2) : 366(1982). Cox & Cox, The Encyclopedia of Rhododendron Species 63 (1997).

Syn.: Series *Fulvum* sensu Tagg in Stevenson (ed.) The Species of Rhododendron 287 (1930).

Large shrubs or small trees, 2 - 10 m high. Bark rough. Branchlets fulvous to greyish-tomentose. Leaves 6 - 27 x 2 - 9 cm, thick, leathery, elliptic to oblong, upper surface with indumentum in young leaves, becoming glabrous on maturity, lower surface with dense unistrate or bistrate granular indumentum, the upper layer in bistrate indumentum is fulvous consisting of capitellate hairs, or greyish and woolly, the lower layer consists of dendroid hairs. Inflorescence 5 - 30-flowered, compact; rachis 5 - 15 mm. Calyx minute. Corolla campanulate to funnel-campanulate, 5-lobed, without nectar pouches, white to pale pink to rose coloured, with a basal blotch. Stamens 10. Ovary glabrous, slender and curved.

Type species : *R. fulvum* Balf.f. & W.W. Smith

Key to the species of subsection Fulva

1.	Leaf indumentum bistrate, the upper layer fulvous, composed of capillate hairs	*fulvum* subsp. *fulvum*
1a.	Leaf indumentum unistrate or bistrate, white or silvery, plastered or woolly, composed of dendroid hairs	*uvarifolium* var. *uvarifolium*

R. fulvum Balf.f. & W.W. Smith, Notes R.B.G. Edinb.10:11(1917) subsp**. fulvum.** Chamberlain, Notes R.B.G. Edinb. 39(2) : 366(1982). Cox & Cox, The Encyclopedia of Rhododendron Species 63 (1997).

Shrub or small tree, 3 - 9 m high. Young shoots rufous-tomentose. Leaves 10 - 21 x 2 - 9 cm, oblanceolate to elliptic, apex rounded, apiculate, base cuneate, upper surface dark green and shiny, lower surface finely granular in appearance, indumentum bistrate, dense, cinnamon, the upper layer with fulvous capitellate hairs, the lower with white stellate-velutinous hairs; petioles 1 - 2 cm, with compacted tomentum intermixed with stipitate glands. Inflorescence 10 - 20-flowered; pedicels 2 - 3 cm, slender, glabrous. Calyx small glabrous. Corolla campanulate, 2.6 - 5 cm long, white to pink, usually blotched at base. Ovary glabrous; style glabrous. Capsule 2.5 - 4 cm long, slender, strongly curved.

Distribution and habitat: India : E. Arunachal Pradesh (*Kingdon-Ward* 8300, Assam and Mishmee hills 1927-28 expedition, *Rhododendron Handbook* 88:1963). China (W. Yunnan, S. E. Xizang), N.E. Burma. In Rhododendron thickets and deciduous forests in between 2,500 - 4,000 m altitude **Map 19.**

Notes: The species derives its name from the tawny tomentum covering the branchlets.

R. uvarifolium Diels, Notes R.B.G. Edinb. 5:213 (1912) var. **uvarifolium** Chamberlain, Notes R.B.G. Edinb. 39(2) : 368(1982). Cox & Cox, The Encyclopedia of Rhododendron Species 66 (1997).

Shrub or small tree, 2 - 10 m high with beautiful silvery, soft young growth. Branchlets white to grey tomentose. Bark rough, greyish to brown. Leaves 7 - 27 x 2.2 - 7 cm, oblanceolate to obovate or oblong, apex rounded to acute or acuminate, base rounded to cuneate, upper surface semi-rugulose, glabrous when mature, lower surface with unistrate or bistrate, usually white to ash-grey, sometimes fawn indumentum, upper layer (when present) composed of dendroid hairs, the lower layer compacted, also with scattered stipitate glands; petioles 1 - 2.5 cm with a whitish tomentum interspersed with a few glands. Inflorescence 6 - 20 or more-flowered (sometimes up to 30); pedicels 2 - 3 cm, sparsely dendroid-tomentose and glandular. Calyx minute, *ca* 1 mm, glabrous. Corolla campanulate to funnel-campanulate, 2.7 - 4 cm long, white, white flushed with rose, usually spotted crimson and with a crimson blotch at the base. Stamens 10. Ovary glabrous; style glabrous. Capsule up to 4.5 cm long, strongly curved.

Distribution and habitat: India : E. Arunachal Pradesh. China (N.W. Yunnan, S.W. Sichuan & S.E. Xizang). In Rhododendron thickets, conifer and mixed forests in between 2,100 - 4,300 m altitude (*Kingdon-Ward* 19457, Lohit valley / Tibet Frontier expedition, 1950; *Rhododendron Handbook*,1963). **Map 19.**

Notes : The name *uvarifolium* is indicative of the *Uvaria*-like leaves.

Subsection Glischra

Subsection **Glischra** (Tagg) Chamberlain, Notes R.B.G. Edinb. 36:116 (1978) & Notes R.B.G. Edinb. 39(2) : 283(1982). Cox & Cox, The Encyclopedia of Rhododendron Species 67 (1997).

Syn.: Series *Barbatum* Tagg subseries *Glischrum* Tagg in Stevenson (ed.) The Species of Rhododendron 137 (1930).

Bushy shrubs or small trees, 1 - 9 m high. Bark rough, grey or light brown. Young shoots often setose-glandular. Leaf bud scales often persistent. Leaves ovate to oblanceolate, upper surface glabrous or with persistent bristles, lower surface with stipitate glands or bristles or with a dense matt of indumentum of ramiform hairs. Inflorescence 6 - 20-flowered, lax. Calyx 5 - 15 mm, lobes rounded, ligulate. Corolla campanulate to funnel-campanulate, 5-lobed, without nectar pouches, white to pink or rose-purple, often with a purple basal blotch and spots. Stamens 10. Ovary densely stipitate-glandular; style glabrous or glandular at base.

Type species : *R. glischrum* Balf.f. & W.W. Smith

Key to the species of subsection Glischra

1. Leaf upper surface very shiny, dark green, sometimes bullate, lower surface with a dense covering of loosely matted indumentum of branched (ramiform) hairs with sparse stipitate glands .... *crinigerum* var. *crinigerum*

1a. Leaf upper surface bright green, with a sparse covering of persistent setae even on older leaves, lower surface densely covered with hispid, long, curved setae .... *glischrum* subsp. *rude*

R. crinigerum Franchet, J. Bot. (Morot) 12:200 (1908) var. **crinigerum** Chamberlain, Notes R.B.G. Edinb. 39(2) : 288(1982). Cox & Cox, The Encyclopedia of Rhododendron Species 68 (1997).

Syn.: *R. ixeunticum* Balf.f. & W.W. Smith, Notes R.B.G. Edinb. 9:240 (1916).

Shrub or small tree, 1 - 5 m high. Young shoots stipitate glandular. Leaf bud scales persistent. Leaves 7.5 - 20 x 2 - 5 cm, lanceolate or oblong-oblanceolate or elliptic or variable, upper surface glossy, dark green and slightly bullate, lower surface with a densely matted white to deep buff indumentum consisting of ramiform (banched) hairs and sparse stipitate glands; petiole glandular. Inflorescence 10 - 20-flowered; pedicels glandular and sparsely hairy, 2.5 - 3 cm long. Calyx 5 - 10 mm, densely stipitate-glandular, lobes irregular, ligulate, rounded. Corolla campanulate, 3 - 4 cm long, white, flushed rose or pink with an intense dark red basal blotch. Ovary stipitate-glandular; style glabrous or glandular at base. Capsule *ca* 15 x 6 mm.

Distribution and habitat: India : Lohit valley, Delei valley in Eastern Arunachal Pradesh adjoining S.E. Xizang. N. Burma, China (Yunnan & S.E. Xizang). Rare. In open conifer forests, among bamboo, boulders and on cliffs in the altitudes of 2,400 - 4,300 m. **Map 19.**

Notes: This species is known only from Kingdon-Ward's collections (No. 8164 – Delei valley & No. 19449-Lohit valley) respectively made during Assam, Mishmee Hills expeditions in 1927 - 1928 and Lohit valley / Tibet frontier expedition in 1950.

The specific epithet *crinigerum* refers to the hairy character on the lower surface of the leaves.

R. glischrum Balf.f. & W.W. Smith, Notes R.B.G. Edinb. 9:229 (1916) subsp. **rude** (Tagg & Forrest) Chamberlain, Notes R.B.G. Edinb. 36:117 (1978) and Notes R.B.G. Edinb. 39(2) : 286(1982). Cox & Cox, The Encyclopedia of Rhododendron Species 70 (1997).

Syn.: *R. rude* Tagg & Forrest in Stevenson (ed.) The Species of Rhododendron 141(1930).

Shrub, up to 3 m high. Branchlets bristly-glandular. Leaves and flower buds sticky. Leaves 11.5 - 30 x 3 - 8 cm, obovate – oblanceolate, apex cuspidate, base rounded, bristly-glandular on both the surfaces; petiole 1 - 2 cm, densely glandular-setose. Inflorescence *ca* 10-flowered; pedicels densely stipitate-glandular. Calyx 5 - 10 mm, reddish-tinged, lobes ciliate on the margins, persistent. Corolla openly campanulate, 2.5 - 3 cm long, pink to purplish-crimson with darker lines. Ovary densely glandular; style glabrous or glandular at base. Ovary stipitate-glandular.

Distribution and habitat: India : N.E. Arunachal Pradesh in Lohit valley. China (N.W. Yunnan, S.E. Xiznag). **Map 19.**

This subspecies is included on the authority of Mr. P.A. Cox (pers. commun. to A.R.K.S).

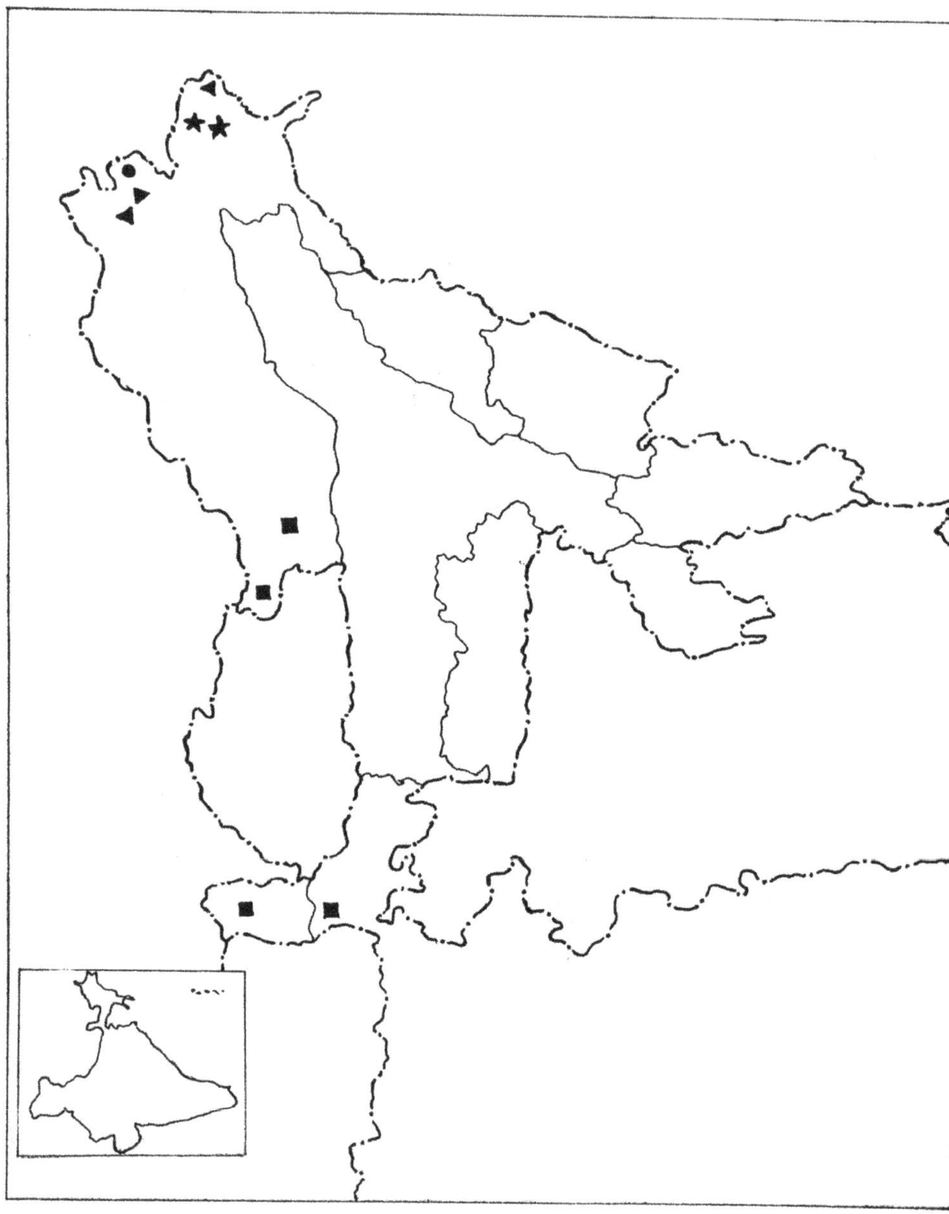

Map 19. ■ R.fulgens. ● R.fulvum ssp.fulvum. ★ R.uvarifolium var.uvarifolium. ▲ R.crinigerum var.crinigerum. ▼ R.glischrum ssp.rude.

(Maps are only indicative and not to scale)

Notes: Cox & Cox (*The Encyclopedia of Rhododendron Species* 71(1997) give a photograph of this subspecies namely *R. glischrum* subsp. *rude* in cultivation at Glenarn, W. Scotland stating it to have been certainly raised from Kingdon-Ward collection No. 10952 collected during his Assam and upper Burma expedition in 1933. However, this collection number has been listed as *R. vesiculiferum* in the *Rhododendron Handbook* 219 (1963) *Part-1*, published by the Royal Horticultural Society, London, and is very closely allied to *R. glischrum.*

The specific epithet *glischrum* means sticky, referring to its sticky branchlets, leaves and flower buds.

Subsection Grandia

Subsection **Grandia** Sleumer, Bot. Jahrb. 74:549 (1949). Chamberlain, Notes R.B.G. Edinb. 39(2) : 241(1982). Cox & Cox, The Encyclopedia of Rhododendron Species 74 (1997).

Syn.: Series G*rande* sensu Tagg in Stevenson (ed.) The Species of Rhododendron 305 (1930).

Large shrubs or trees, up to 30 m. Bark rough; branchlets thick, glabrous to tomentose. Leaves to 70 x 30 cm, oblanceolate to broadly elliptic, glabrous on the upper surface (on maturity), covered with unistrate or bistrate, plastured indumentumon the lower surface. Inflorescence 10 - 30-flowered; rachis long, 2 - 6 cm. Calyx minute, 1 - 2 mm. Corolla fleshy, 6 - 10-lobed, tubular-campanulate or campanulate, oblique or ventricose, usually without nectar pouches, white, yellow or pink purple. Stamens 12 - 20. Ovary tomentose, sometimes glandular or glabrous; style glabrous.

Type species : *R. grande* Wight

This subsection has some of the tallest and robust tree species of the genus having the largest leaves and inflorescences.

Key to the species of subsection Grandia

1.	Corolla ventricose or oblique-campanulate; leaf indumentum silvery, compacted and agglutinated, unistrate........................	2
1a.	Corolla tubular or funnel-campanulate; leaf indumentum absent or silvery to buff-tomentose, not compacted and not agglutinated, bistrate, sometimes unistrate	4

2. Ovary and pedicels glandular (not tomentose) .... *grande*

2a. Ovary and pedicels tomentose (not glandular) 3

3. Leaves 8 - 28 cm wide, 2.2 - 2.8 x as long as broad (2 - 3 times longer than broad); corolla pale creamy white; stamens 18.. ...... *sinogrande*

3a. Leaves 3 - 7 cm wide, (2.5) 3 - 5 x as long as broad (3 - 5 times longer than broad); corolla yellow; stamens 16 .... *sidereum*

4. Corolla lemon-yellow; leaf indumentum bistrate, with a lanate-tomentose upper layer .... *macabeanum*

4a. Corolla pink to rosy purple ; leaf indumentum unistrate or bistrate, not lanate-tomentose 5

5. Pedicels 20 - 25 mm; leaf indumentum a buff tomentum, sometimes restricted to the leaf margins, or more or less absent .... *prostitum*

5a. Pedicels 8 - 15 mm; leaves with a whitish, felted or arachnoid indumentum .... *wattii*

R. grande Wight, Calcutta J. Nat. Hist. 8 : 176 (1874) & Ic. Ind. Pl. Orient.t.1202 (1850). C.B. Clarke in Hook.f., Fl. Brit. India 3:464 (1882). Chamberlain, Notes R.B.G. Edinb. 39 (2) : 244 (1982). Cox & Cox, The Encyclopedia of Rhododendron Species 75(1997).

Syn.: *R. argenteum* Hook.f., Rhododendrons of Sikkim Himalaya t. 9 (1849).

Tree, 6 - 15 m high with a well-developed trunk. Branchlets whitish tomentose. Leaves evergreen 15 - 46 x 8 - 13 cm, elliptic to oblanceolate, upper surface deep green, shining and glabrous or rugulose with the main nerves strongly impressed, lower surface covered with a thin, compacted, woolly, or fawn silvery indumentum; petioles 2 - 3.5 cm long, sparsely floccose becoming glabrous on ageing. Inflorescence 15 - 25-flowered; rachis 4 - 5 cm long; pedicels 1.5 - 3 cm, densely covered with sessile glands. Calyx *ca* 1 mm, glandular. Corolla 8-lobed, ventricose-campanulate, 5.7 cm long, cream, pale yellow to light pink with a blotch and spots, nectar pouches 8, purple coloured. Stamens 15 - 16. Ovary densely glandular, sometimes with pale brown tomentum; stigma large. Capsule 3 - 4.5 cm long, tomentose, curved.

Distribution and habitat: India : Darjeeling Hills, West Bengal, Sikkim, West & Central Arunachal Pradesh. E. Nepal, Bhutan, China (S. Xizang). Fairly common in wet evergreen and mixed forests growing in association with other *Rhododendron* species and Magnoliaceae members, in the altitudinal range of 2,100 - 3,200 m. **Map 20.**

Notes: The species name *grande* refers to its large and showy leaves and inflorescence.

R. macabeanum Watt ex Balf.f., Notes R.B.G. Edinb. 12:128 (1920). Chamberlain, Notes R.B.G. Edinb. 39(2) : 251(1982). Cox & Cox, The Encyclopedia of Rhododendron Species 77 (1997).

Syn.: *R. falconeri* Hook.f. var. *macabeanum* Watt, mss.

R.grande Wight

Erect tree to 15 m high. Young shoots whitish woolly-tomentose; leaf bud scales scarlet red. Leaves 14 - 38 x 9 - 20 cm, broadly ovate to broadly elliptic, apex rounded to retuse, sometimes minutely apiculate, base cuneate, upper surface dark green, shiny with impressed main veins and often with vestiges of indumentum, lower surface with a dense, bistrate indumentum, the lower layer compacted, the upper layer with lanate-tomentose, rosulate and ramiform hairs; petioles 2 - 2.5 cm, slightly winged, floccose-tomentose. Inflorescence compact, 15 - 25 (- 30)-flowered ; rachis up to 3.5 cm ; pedicels 2.5 - 4 cm, densely whitish-felted tomentose. Calyx *ca* 1 mm, tomentose, lobes triangular. Corolla 8 - lobed, tubular to funnel-campanulate, 5 - 7.5 cm long, lemon yellow with a purplish blotch in the throat. Stamens 16. Ovary rufous-tomentose. Capsule 2 - 4 cm long, curved.

Distribution and habitat: India : Manipur and Nagaland. Endemic, forming dense stands on the summit of hills and scattered in mixed forests in the altitudes of 2,500 - 3,000 m associated with birch. **Map 20.**

Notes: The species is named after Mr. McCabe, a former British Deputy Commissioner of Naga Hills in North East India.

R. prostitum Balf.f. & Forrest, Notes R.B.G. Edinb. 12:131 (1920) var. **prostitum.** Chamberlain, Notes R.B.G. Edinb. 39(2) : 248(1982). Cox & Cox, The Encyclopedia of Rhododendron Species 81 (1997).

Syn.: *R. giganteum* Tagg var. *seminudum* Tagg & Forrest, Notes R.B.G. Edinb. 15:108 (1926).

Tree, 6 - 10 m high. Branchlets thin, greyish-tomentose. Leaf bud scales bright red. Leaves 20 - 56 x 9 - 25 cm, ellitptic to oblanceolate, apex rounded, retuse or minutely apiculate, base nearly rounded, upper surface matt green, rugulose with the main veins impressed, lower surface glabrous in juvenile state, developing a buff, continuous, thin indumentum at least along a marginal band as the plant matures; petioles 2 - 3 cm, terete, glabrous. Inflorescence 20 - 30-flowered; rachis 2 - 3 cm, thick ; pedicels 20 - 25 mm, densely reddish-brown tomentose. Calyx *ca* 2 mm, tomentose. Corolla funnel-campanulate, 8-lobed, 6 - 7.5 cm long, pink to rosy purple, sometimes two-tone, fading, with a dark basal blotch and nectar pouches. Stamens 16. Ovary densely rufous-tomentose. Capsule 4 - 5 x 1.5 cm.

Distribution and habitat: India : Eastern Arunachal Pradesh. N.E. Upper Burma, China (W. & N.W. Yunnan, S.E. Xizang). Rare. In dense mixed forests with other species of *Rhododendron* at altitudes ranging from 2,700 - 4,000 m. **Map 20.**

Notes : The species name *prostitum* means first of the first.

R. sidereum Balf.f., Notes R.B.G. Edinb. 12:162 (1920). Chamberlain, Notes R.B.G. Edinb. 39(2) : 246(1982). Cox & Cox, The Encyclopedia of Rhododendron Species 83 (1997).

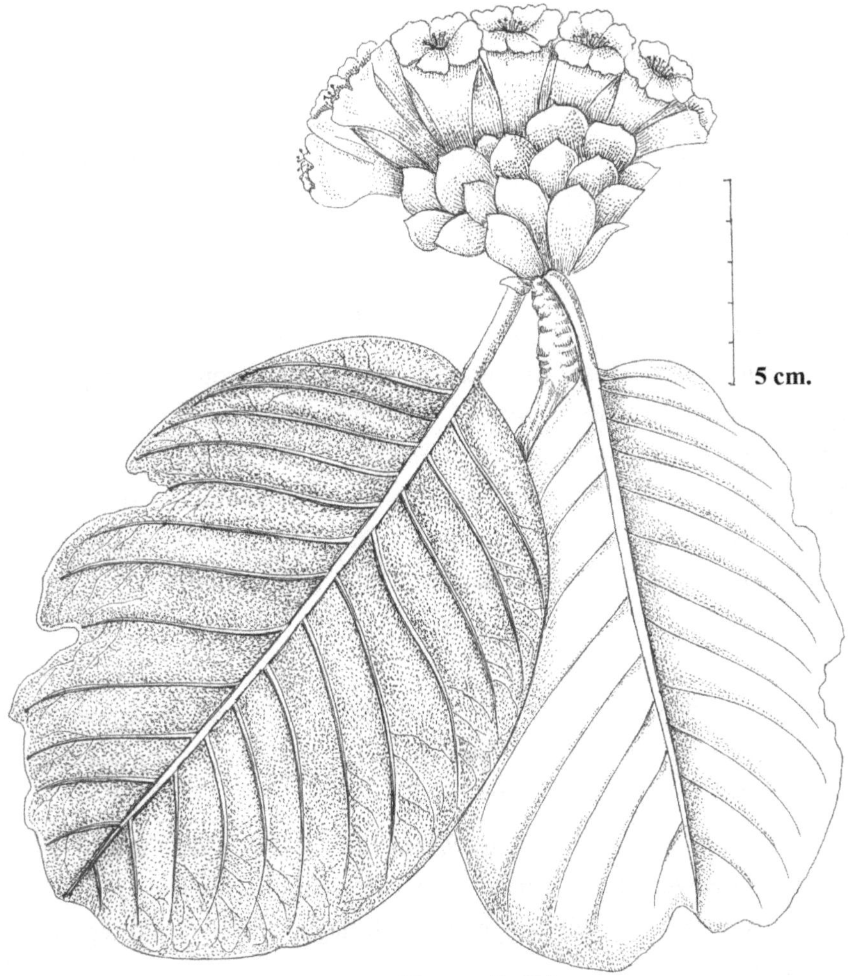

R.macabeanum Watt ex Balf.f.

Large shrub or small tree, to 12 m high. Branchlets slender, tomentose. Leaves (9 -) 16 - 20 x 4 - 7 cm, oblanceolate to narrowly ellitptic, apex acute to rounded, apiculate, base rounded to attenuate,upper surface dark mat green,lower surface with silvery-white to fawn indumentum; petioles 2 - 2.5 cm, terete, glabrous. Inflorescence 12 - 20-flowered; rachis 50 - 60 cm; pedicels 2.5 - 3.5 cm, brownish tomentose. Calyx rim like, *ca* 1 mm, tomentose. Corolla 8-lobed, ventricose-campanulate, *ca* 5cm long, cream to clear yellow, with a crimson-red basal blotch. Stamens 16. Ovary densely rufous-tomentose. Capsule 25 - 30 cm long, 1 cm thick, straight or curved.

Distribution and habitat: India : Eastern Arunachal Pradesh. Upper Burma, China (N.W. Yunnan); mostly in conifer and mixed forests occurring in the altitudinal range of 2,400 - 3,800 m. **Map 20.**

Notes: The specific epithet s*idereum* means excellent.

R. sinogrande Balf.f. & W.W. Smith, Notes R.B.G. Edinb. 9:274 (1916). Chamberlain, Notes R.B.G. Edinb. 39(2) : 245(1982). Cox & Cox, The Encyclopedia of Rhododendron Species 84(1997).

Tree, 14 m high. Branchlets thick, sparsely pale tomentose. Leaves very large, beautiful, 20 - 70 (- 90) x 8-30 cm, 2.2 - 2.8 x as long as broad, oblanceolate to broadly elliptic, apex rounded or retuse, minutely apiculate, base rounded, upper surface dark green, shiny and rugulose with impressed veins, lower surface with a plastered silvery-grey or fawn, shiny indumentum; petiole rounded, 3 - 5 cm, with a silvery agglutinated (plastured) indumentum. Inflorescence 12 - 30-flowered; rachis 3 - 6.5 cm; pedicels 3-6 cm, densely villous; tomentose. Calyx *ca* 2 mm, lobes minute, triangular, tomentose. Corolla 8 - 10-lobed, ventricose-campanulate, 6 - 7.5 cm long, creamy-white to pale yellow with a crimson blotch at the base. Stamens 18 - 20. Ovary densely rufous-tomentose.

Distribution and habitat: India : Western Arunchal Pradesh-Kameng & Tawang Districts. N. Burma, China (N.W. Yunnan, S.E. Xizang). In mixed forests, in the altitudinal range of 2100 - 3400 m. **Map 20.**

Notes : The largest leaved *Rhododendron* with shiny, dark green upper surface contrasting with the silvery fawn, shiny lower surface rendering easy identification in the wild, this species looks very elegant even in foliage. The specific epithet means the 'Chinese *R. grande.'*

R. wattii Cowan, Notes R.B.G. Edinb. 19:163, t. 253 (1936). Chamberlain, Notes R.B.G. Edinb. 39(2) : 242(1982).

Shrub or small tree, to 7 m high. Leaves 11 - 16 x 5 - 7 cm, 2 - 2.3 x as long as broad, obovate to oblong, apex rounded, apiculate, base rounded, upper surface glabrous, lower surface with a sparse whitish felted indumentum, lateral veins prominent; petiole about 2 cm long, terete, glabrous on maturity. Inflorescence *ca* 15-flowered, dense; pedicels about 1 cm long, glandular. Calyx 1 - 2 mm, glandular; lobes broadly triangular. Corolla 6-lobed, tubular-campanulate, pink with darker flecks and purplish basal patches, 3.5 - 5.5 cm. Ovary covered with a densely lanate-pilose, brownish indumentum.

Distribution and habitat: India : Ching Sow, Manipur. Grows on the forested mountain ridges, at an altitude of 2700 m. Endemic and endangered and is so far known only from its Type locality. **Map 20.**

Notes: The species is named after George Watt, the Imperial Economic Botanist, who collected plants in Manipur and Nagaland and who authored the most valuable Watt's Dictionary of Economic Products of India.

R. kesangiae D.G. Long & Rushforth

Shrub or small tree, 8 - 12 m high. Branchlets often covered with tufts of soft hair. Bark brown, not peeling. Terminal buds rounded, green, reddish to nearly black. Leaves 20 - 30 x 10 - 16 cm, broadly elliptic to nearly obovate, apex and base usually rounded, upper surface glabrous with prominent lateral veins, lower surface with woolly to almost plastered, fawn to silvery-white indumentum on young leaves, without cup-shaped hairs, becoming uniformly grey-white or sometimes brownish-fawn on old leaves; petiole 2 - 3.5 cm long, not winged. Inflorescence 15 - 20-flowered, compact. Calyx small. Corolla campanulate, with nectar pouches, 3 - 4.7 cm long, rose to pink, fading to pale pink or white. Ovary densely glandular, sometimes thinly tomentose.

Distribution and habitat: India: W.Arunachal Pradesh. Bhutan. In temperate rhododendron forests associated with fir, hemlock or bamboo, in the altitudes of 2,900 - 3,500 m.

Notes: This species is intermediate between *R. hodgsonii* and *R. falconeri,* hybridizing with both the species in the wild. Two varieties, namely, *kesangiae* (with rose to pink flowers) and *album* Namgyel & D.G. Long (with white flowers) are distinguished.

According to Mr. P.A. Cox, (Pers. Comm. to A.R.K.S.) the species occurs in Arunachal Pradesh (Tawang and Kameng districts). It is named in honour of the Queen Mother of Bhutan.

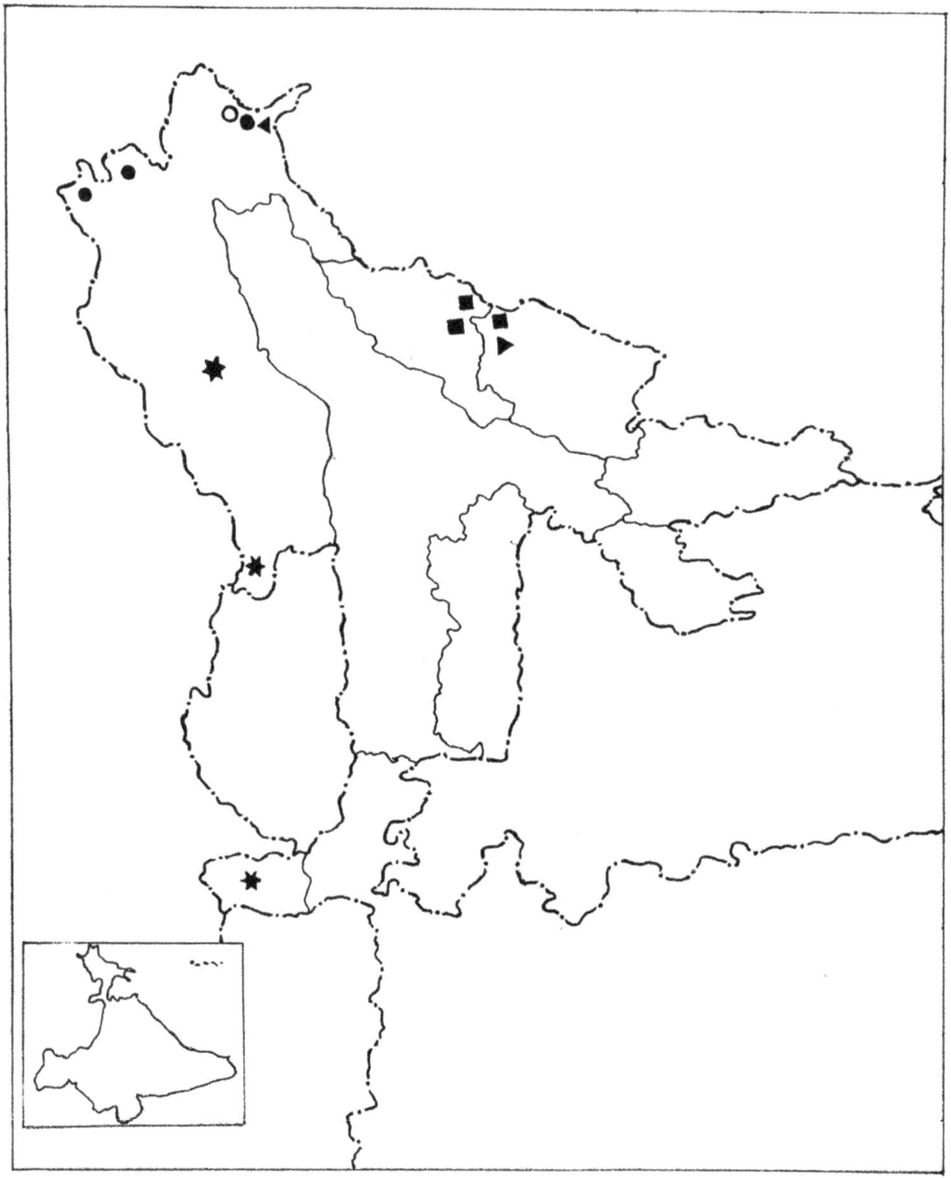

Map 20. ★ R.grande. ■ R.macabeanum. ▲ R.prostitum var.prostitum O R.sidereum. ● R.sinogrande. ▼ R.wattii

(Maps are only indicative and not to scale)

Subsection Irrorata

Subsection **Irrorata** Sleumer, Bot. Jahrb. 74:548 (1949). Chamberlain, Notes R.B.G. Edinb. 39(2) : 290(1982). Cox & Cox, The Encyclopedia of Rhododendron Species 86 (1997).

Syn.: Series *Irroratum* subseries *Irroratum* sensu Tagg in Stevenson (ed.) The Species of Rhododendron 331 (1930).

Shrubs or small trees, 1 - 12 m high. Bark rough. Young shoots stipitate-glandular, sometimes also with a floccose tomentum. Leaves to 19 cm long, 6.5 cm broad, ovate to oblanceolate, elliptic or oblong, upper surface tomentose when young becoming glabrous on maturity, lower surface glabrous on maturity with punctate bases or thinly covered with dendroid hairy indumentum, glandular on the veins. Inflorescence 4 - 20-flowered, lax or dense; rachis 5 - 10 (- 35) mm. Calyx minute or cup-shaped. Corolla tubular to open campanulate, 5 - 7-lobed, usually with nectar pouches, white to mauve or deep crimson, often with darker flecks and also with a basal blotch. Stamens 10. Ovary glabrous, tomentose or glandular; style glabrous or glandular to tip.

Type species : *R. irroratum* Franchet

Key to the species of subsection Irrorata

1. Leaves narrowly elliptic to lanceolate-acuminate, 4 - 6 times longer than broad, margins *strongly* undulate; inflorescence 10 - 20-flowered; pedicels sparsely stellate-tomentose .... *kendrickii*

1a. Leaves elliptic to oblanceolate or oblong, 2.5 - 5 times longer than broad, margins *not* undulate; inflorescence 4 - 10-flowered; pedicels densely dendroid-tomentose or tomentose with or without glands 2

2 Corolla campanulate, nectar pouches absent, pale cream or pink coloured; pedicels dendroid-tomentose .... *papillatum*

2a. Corolla tubular-campanulate with well-developed nectar pouches, crimson coloured; pedicels tomentose with or without glands .... *tanastylum*

R. kendrickii Nuttall, J. Bot. (London) 5:358 (1853). Chamberlain, Notes R.B.G. Edinb. 39(2) : 302(1982). Cox & Cox, The Encyclopedia of Rhododendron Species 91(1997). **(Photos No. 69).**

Loosely branched shrub or small tree, 3 - 8 m high. Branchlets slender. Leaves 9 - 13.5 x 2 - 3 cm, narrowly elliptic-lanceolate to lanceolate-acuminate, sub-coriaceous, cuneate to rounded at base, margins strongly undulate, upper surface smooth, lower surface paler and shiny, hairy along the midrib; petioles 1 - 1.5 cm, glabrescent. Inflorescence dense, 10 - 20-flowered; pedicels *ca* 8 mm, sparsely stellate-tomentose. Calyx 2 - 3 mm, lobes well-developed, rounded, glabrous or sparsely glandular-ciliate. Corolla thick, tubular-campanulate, 3 - 4 cm long, with 5 conspicuous nectar pouches at base, deep rose to scarlet, with darker flecks; pedicel *ca* 8 mm long, sparsely stellate-tomentose. Ovary with a few rufous dendroid hairs; style glabrous. Capsule 1.5 - 2.5 cm long, curved.

Distribution and habitat: India : Arunachal Pradesh. Bhutan, China (S. Xizang). Frequent in dense, mixed evergreen forests in the altitudes of 2,100 - 2,700 m. **Map 21.**

Notes: This species is easily identified in the field by its stiff, narrowly lanceolate leaves with pointed apices and characteristically wavy margins and dense trusses of deep rose or scarlet red flowers. The species is named after Dr. Kendrick, a friend of Dr. Thomas Nuttall who collected plants in Bhutan and western Arunachal Pradesh.

R. papillatum Balf.f. & Cooper, Notes R.B.G. Edinb. 13:282 (1922). Chamberlain, Notes R.B.G. Edinb. 39(2) : 304(1982).

Syn.: *R. epapillatum* Balf.f. & Cooper, Notes R.B.G. Edinb. 13:257 (1922).

Shrub or small tree, 2 - 5 m high. Leaves 9 - 14 x 3 - 5 cm, *ca* 3 x as long as broad, oblanceolate to oblong, sub-coriaceous, base rounded to cuneate, margins not undulate, upper surface glabrous on maturity, lower surface usually with conspicuous veins and a papillate cuticle and with a sparse stellate indumentum; petioles 1.5 - 2 cm, glabrous. Inflorescence 5 - 10-flowered; pedicels *ca* 2 cm long, densely dendroid-tomentose. Calyx *ca* 2 mm, tomentose. Corolla campanulate, 4 - 5.5 cm long, without nectar pouches, pale cream to pink, with purple flecks and a basal blotch. Ovary densely dendroid-tomentose and stipitate glandular.

Distribution and habitat: India : Arunachal Pradesh. Bhutan. Rather rare; in *Abies* forests as undergrowth, in the altitudinal range of 1800 - 3300 m. **Map 21.**

Notes: The species name *papillatum* refers to the nipple-like projections of the cuticle on the undersurface of the leaf.

R. tanastylum Balf.f. & Kingdon-Ward, Trans. Bot. Soc. Edinb. 27:217 (1917) var. **tanastylum.** Chamberlain, Notes R.B.G. Edinb. 39(2):300(1982). Cox & Cox, The Encyclopaedia of Rhododendron Species 83(1997).

Syn.: *R. cerochitum* Balf.f. & Forrest, Notes R.B.G. Edinb. 13:36 (1920).

Shrub or small tree, up to 9 m high often straggly in habit. Branchlets slender, thinly floccose at first. Leaves 7.5 - 15 x 3 - 6 cm, 2.5 - 5 x as long as broad, elliptic to oblanceolate, leathery, base cuneate, margins not wavy, lower surface with a thin veil of indumentum, becoming glabrous on maturity; petiole 1 - 2 cm, glabrous. Inflorescence loose, 4 - 8-flowered; pedicels without glands. Corolla tubular-campanulate, with 5 nectar pouches at the base, 4.5 - 5.5 cm long, deep pink to purple-crimson with deeper spots. Ovary glabrous to tomentose, glandular; style glabrous. Capsule 11 - 23 x 5 - 8 mm.

Distribution and habitat: India : N.E. Arunachal Pradesh. N.E. Burma, China (W. Yunnan). Common; in broad-leaved mixed pine forests and *Rhododendron* thickets, in the altitudinal range of 1800 - 3400 m. **Map 21.**

Notes: The species derives its name from its long style (*tanastylum* = long styled).

Subsection Lanata

Subsection **Lanata** Chamberlain, Notes R.B.G. Edinb. 37:337 (1979) & Notes R.B.G. Edinb. 39(2) : 368(1982). Cox & Cox, The Encyclopedia of Rhododendron Species 94 (1997).

Syn.: Series *Campanulatum* sensu Tagg, pro parte. Subsection *Campanulata* Sleumer, pro parte.

Shrub, 0.3 - 4 m high. Young shoots and buds densely lanate-tomentose. Bark greyish-purple, slightly rough. Leaves 3.5 - 11 cm long, obovate to elliptic, upper surface often with persistent indumentum, lower surface with a dense, unistrate, light brown to rufous (rarely whitish when young) lanate or crisped indumentum composed of dendroid hairs. Inflorescence 3 - 15-flowered. Calyx minute. Corolla campanulate or open-campanulate, without nectar pouches, yellow, white or pink, crimson-spotted. Stamens 10. Ovary densely tomentose; style glabrous. Capsule tomentose.

Type species : *R. lanatum* Hook.f.

Key to the species of subsection Lanata

1.	Leaves on the upper surface with persistent indumentum	*tsariense*
1a.	Leaves on the upper surface pale green, matt, hairy especially in the grooved midrib or with reddish brown indumentum which wears off on maturity	2
2.	Leaves oblong-elliptic or elliptic, thin in texture, upper surface pale green, matt, hairy especially in the grooved midrib, lower surface with a thin, felted, rust-brown indumentum; corolla pale yellow to pale pink, crimson spotted	*flinckii*
2a.	Leaves narrowly-elliptic, obovate to oblong-obovate, thick and leathery in texture, upper surface with reddish-brown indumentum which wears off on maturity, lower surface with crisped, thick, grey-fawn, pink-brown, or rust-coloured indumentum; corolla cream to sulphur yellow, red or brown spotted	*lanatum*

R. flinckii Davidian, Quart. Bull. Amer. Rhododendron Soc. 25:143 (1975). Chamberlain, Notes R.B.G. Edinb. 39(2) : 369(1982) as synonym under *R. lanatum.* Cox & Cox, The Encyclopedia of Rhododendron Species 95 (1997).

Shrub, 1.5 - 2.5 m high. Branchlets densely tomentose with rusty brown, fawn or cottony indumentum. Leaves 4-10 x 2 - 4.5 cm, oblong-oval, oblong-elliptic or elliptic, thin in texture, upper surface pale green, matt, hairy or with vestiges of hairs in the midrib groove, lower surface with a thin, felted, bright rust-brown, continuous, unistrate indumentum; petiole densely tomentose. Inflorescence 3 - 8-flowered. Calyx minute. Corolla campanulate, 3.5 - 5 cm long, pale yellow to pale-pink with crimson spottings. Ovary tomentose; style glabrous.

Distribution and habitat: India : W. Arunachal Pradesh. Bhutan. In *Abies* forest, forest margins and on cliffs, in the altitudinal range of 3,000 - 4,150 m. **Map 21.**

Notes: Chamberlain (loc. cit) reduced this species to synonymy under *R. lanatum* to which it is closely related. Cox & Cox treat *R. flinckii* as a separate species distinctive enough from *R. lanatum* which has thicker leaves, shiny and dark green rather than matt on the upper surface, with a thick, dense, indumentum on the lower surface, and state that both the species merge in the wild. This species (*R. flinckii)* is included in this book on the authority of Mr. Peter A. Cox (pers. comm. to A.R.K.S).

R. lanatum Hook.f. Rhododendrons of Sikkim Himalaya 17, t. 16 (1849). C.B. Clarke in Hook.f., Fl. Brit. India 3:467 (1882). Chamberlain, Notes R.B.G. Edinb. 39(2) : 369(1982). Cox & Cox, The Encyclopedia of Rhododendron Species 96 (1997). **(Photo Nos. 70, 70a).**

Bushy shrub, up to 4 m high. Young shoots densely clothed with soft adpressed white to tawny cottony tomentum. Leaf and flower buds heavily tomentose. Leaves tending to be confined to the terminal portions of the branchlets, 6 - 12 x 1.8 - 5 cm, narrowly elliptic or obovate to oblong-obovate, thick and leathery, apex rounded, shortly apiculate, base rounded, upper surface with reddish-brown, floccose indumentum which wears off on maturity, persisting on the midrib, lower surface with a dense, crisped unistrate indumentum of dendroid hairs, whitish when young, becoming grey-fawn, pink-brown, brown or rust-coloured on ageing; petioles *ca* 1 cm, densely tomentose. Inflorescence 5 - 10-flowered; pedicels 1 - 2 cm, densely white to brown tomentose. Calyx *ca* 1 mm, tomentose. Corolla broadly-campanulate, 3 - 5 cm long, cream to pale to sulphur yellow, spotted with red or brown. Ovary densely rufous-tomentose; style glabrous. Capsule 15 - 25 x 4 - 8 mm, curved.

Distribution and habitat: India : Sikkim, Arunachal Pradesh. Bhutan, China (S. Xizang). In *Abies* forests, amidst rhododendrons and bamboo, and on cliffs, in the altitudinal range of 3,000 - 4,500 m. **Map 21.**

Notes: The species name *lanatum* refers to the dense, woolly indumentum that covers the branchlets, leaves and flower buds.

R. tsariense Cowan, Notes R.B.G. Edinb. 19:232 (1937). Chamberlain, Notes R.B.G. Edinb. 39(2) : 370(1982). Cox & Cox, The Encyclopedia of Rhododendron Species 97 (1997).

Shrub, 1 - 3 m high. Branchlets densely tomentose. Leaves 2.3 - 6.2 x 1 - 3 cm, obovate to oblong, apex rounded or acute, apiculate, base rounded, upper surface glabrous except for the persistent indumentum on the midrib, lower surface densely rufous-tomentose with ramiform hairs; petioles *ca* 1 cm long, densely tomentose. Inflorescence 3 - 5-flowered; pedicels *ca* 1 cm long, densely tomentose. Calyx *ca* 2 mm, densely tomentose. Corolla open-campanulate, 2.5 - 3.5 cm long, white or pale cream, pale-pink or white flushed with pink, crimson spotted. Ovary densely tomentose. Capsule *ca* 12 x 4 mm.

Distribution and habitat: India : Western Arunachal Pradesh in Tawang and Kameng districts adjacent to Bhutan. E. Bhutan, China (S. Xizang). In conifer and *Rhododendron* forests, mixed with bamboo, along open hill sides and on cliffs in between 3,000 - 4,500 m alt. **Map 21.**

Notes: This species is named after its Type locality Tsari Chu in S. Xizang.

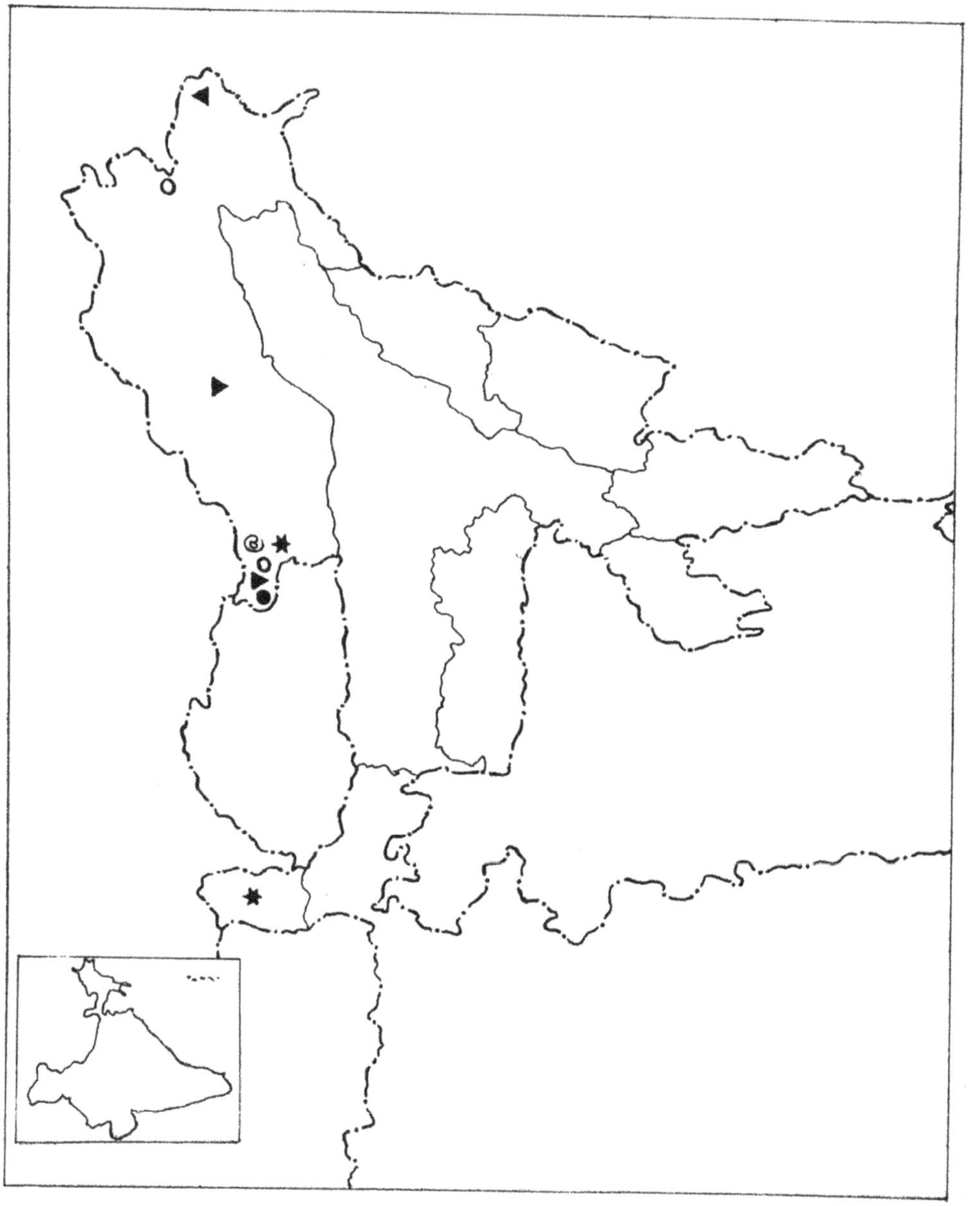

Map 21. ▼ R.kendrickii. ● R.papillatum. ▲ R.tanastylum var.tanastylum. @ R.flinckii. ★ R.lanatum. O R.tsariense.

(Maps are only indicative and not to scale)

Subsection Neriiflora

Subsection **Neriiflora** Sleumer, Bot. Jahrb. 4:545 (1949). Chamberlain, Notes R.B.G. Edinb. 39(2) : 384(1982). Cox & Cox, The Encyclopedia of Rhododendron Species 107 (1997).

Syn.: Series *Neriiflorum* sensu Tagg in Stevenson (ed.) The Species of Rhododendron 507 (1930).

Dwarf or prostrate shrubs or small trees, 0.5 - 6 m high. Branchlets slender, with a thin or sometimes thick, whitish to rufous indumentum consisting of rosulate, dendroid or branching hairs, sometimes with bristles and / or glands. Bark smooth or rough, often peeling, fawn to cinnamon coloured. Leaves narrowly elliptic to orbicular, upper surface glabrous on maturity, lower surface glabrous, or with some indumentum, indumentum continuous or patchy, compact or woolly, whitish to buff or brown or rufous. Inflorescence 1 - 12(- 20)-flowered, lax or dense usually with drooping flowers in between the leaves. Calyx minute to well-developed, cup-like, often coloured, up to 15 (- 20) mm long. Corolla 5-lobed, tubular-campanulate, rarely campanulate, fleshy, with depressed nectar pouches, mostly red, also white, yellow, or with different shades of pink to carmine. Stamens 10. Ovary usually tomentose, with or without stipitate glands, or glabrous; style glabrous.

Type species: *R. neriiflorum* Franchet

A complex subsection with as many as twenty five species mostly growing in a common distribution range (in India and outside) and thereby merging and hybridizing freely in the wild. Cox & Cox (*The Encyclopedia of Rhododendron species* 109, 1997) after considerable field observations on the species and subspecific taxa of some of the species of this Subsection consider that many of the so-called species, subspecies and varieties to be simply extreme forms of the species or natural hybrids. The treatment followed here is in accordance with the treatment given by Cox & Cox in their book who have divided this subsection into alliances corresponding to the old subseries divisions, for easy classification of species in India. About nine species are reported to be distributed in India, though some are of doubtful occurrence, but are included here.

Alliances in the subsection Neriiflora

1.	Bushy or straggling shrubs, 1 - 3 m high; leaves with thick woolly indumentum on the lower surface; corolla red; ovary usually tomentose	Hematodes *(R. beanianum, piercei, pocophorum)*
2.	Dwarf to low shrubs, 0.6 - 2 m high; leaves not more than 10 cm long; indumentum variable, from absent to thick; inflorescence small and lax; corolla variable in colour.	Sanguineum *(R. eudoxum, trilectorum)*
3.	Prostrate to mound-forming shrubs; leaves glabrous or with a thin indumentum / glands; corolla crimson or rarely pink; inflorescence 1 - 5-flowered	Forrestii *(R. chamaethomsonii, forrestii)*
4.	Upright, loosely branched shrubs, 0.3 - 6 m; leaves glabrous or with a loose woolly indumentum; ovary tapering into the style	Neriiflorum *(R. neriiflorum)*

Key to the species of Hematodes Alliance

1.	Erect shrubs, up to 3 m high; branchlets densely glandular; petioles densely bristly-glandualr or stipitate-glandular, tomentose when young; leaf upper surface bullate or shiny and smooth	2
1a.	Straggling shrubs, 1.2 - 2.5 m high; branchlets tomentose; petioles without bristles; leaf upper surface rugulose and shiny, lower surface with a bistrate woolly, fulvous-tomentose dendroid hairy indumentum	*piercii*
2.	Branchlets densely bristly-glandular; petioles densely bristly and grandular; leaf upper surface dark green, bullate, lower surface with a dense unistrate, rufous indumentum	*beenianum*
2a.	Branchlets densely glandular; petioles stipitate glandular, tomentose when young; leaf upper surface shiny, glabrous, lower surface with a unistrate continuous woolly tomentum..............................	*pocophorum* var. *pocophorum*

R. beanianum Cowan, New Flora & Silva 10:245, f.80 (1938). Chamberlain, Notes R.B.G. Edinb. 39(2) : 386(1982). Cox & Cox, The Encyclopedia of Rhododendron Species 108 (1997).

Shrub up to 3 m, straggling with age. Branchlets bristly (stellate-setose) and stipitate-glandular. Bark greenish-brown to red-brown. Leaves 6 - 9 x 3.2 - 4.4 cm, usually obovate-elliptic, sometimes oblong-elliptic to oblong, apex rounded, apiculate, base nearly rounded, upper surface dark, glabrous, bullate, lower surface with a dense unistrate fulvous tomentum consisting of dendroid hairs; petioles 1.5 - 2 cm long, setose to tomentose. Inflorescence 6 - 10-flowered; pedicels 10 - 15 mm, setulose-tomentose. Calyx *ca* 5 mm, cup-shaped, sparsely tomentose. Corolla fleshy, tubular-campanulate, *ca* 3.5 cm long, scarlet, crimson or rarely pink. Stamens 10. Ovary stellate-tomentose. Capsule 12 x 6 mm, tomentose.

Distribution and habitat: India : E. Arunachal Pradesh. N.E. Upper Burma. In bamboo thickets, in between rocks and along forest clearings and edges, in the altitudes of 2,700 - 3,400 m. **Map 22.**

Notes: The species is named after W.J. Bean, a former Curator, Royal Botanic Gardens, Kew. Very closely allied to *R. piercei* but is easily recognized by the densely bristly and glandular branchlets and petioles. *R. beanianum* is much preferred in cultivation for its elegant foliage and brightly coloured flowers.

R. piercei Davidian, Quart. Bull. Amer. Rhododendron Soc. 30:211 (1976). Chamberlain, Notes R.B.G. Edinb. 39(2) : 387(1982). Cox & Cox, The Encyclopedia of Rhododendron Species 111 (1997).

Syn.: *R. beanianum* Cowan var. *compactum* Cowan, Notes R.B.G. Edinb. 21:148 (1953).

Bushy or straggling shrub, 1.2 - 2.5 m high. Branches spreading. Branchlets tomentose. Bark light brown to brown. Leaves 6 - 11 x 2.7 - 5.2 cm, ovate to elliptic, apex rounded, apiculate, base rounded, upper surface tomentose in young leaves, glabrous, shiny and rugulose at maturity, lower surface with a bistrate indumentum, the upper layer thick, fulvous and dendroid-tomentose, the lower white and adpressed; petioles 1 - 2 cm long, tomentose. Inflorescence 6 - 8-flowered; pedicels 1.2 - 1.5 cm long, tomentose. Calyx 3 - 6 mm, irregular, glabrous. Corolla fleshy, with darker nectar pouches. Ovary densely tomentose. Capsule 1.4 - 1.8 cm long.

Distribution and habitat: India: ?N.E. Arunachal Pradesh bordering S.E. Xizang in China. China (S.E. Xizang). The species is included on the authority of Mr. P.A. Cox (pers. comm. to A.R.K.S). **Map 22.**

R. pocophorum (Balf.f. ex) Tagg, Notes R.B.G. Edinb. 15:316 (1927). var. **pocophorum** Chamberlain, Notes R.B.G. Edinb. 39(2) : 387 (1982). Cox & Cox, The Encyclopedia of Rhododendron Species 112 (1997).

Erect shrub, 1 - 3 m high. Branchlets densely stipitate-glandular. Bark light brown. Leaves rigid, 10 - 16 x 4 - 7 cm, oblong to obovate, apex rounded, apiculate, base cuneate, upper surface glabrous, dark, shiny, lower surface with a thick, brown, unistrate, dendroid, continuous, woolly indumentum; petioles 1 - 2 cm long, thick, glandular-tomentose when young. Inflorescence 10 - 20-flowered, compact; pedicels 7 - 18 mm long, glandular. Calyx 5 - 10 mm, lobes irregular, glandular-ciliate. Corolla fleshy, tubular-campanulate, 3.6 - 5.3 cm long, light crimson or crimson scarlet. Ovary densely glandular. Capsule 2 - 2.5 cm long, 7 - 9 mm thick.

Distribution and habitat: India : N.E. Arunachal Pradesh. China (S. Xizang & N.W. Yunnan). Seems to be rather rare; in thickets, margins of bamboo and conifer forests and open rocky slopes. **Map 22.**

Notes: Kingdon-Ward's collection No. 8289 of 04-06-1928 from Delei valley in N.E. Arunachal Pradesh, made during his 1927 expedition to Assam and Mishmee Hills (det. Prof. D.F. Chamberlain), available in the Kew Herbarium and noted by the senior author (A.R.K.S) authenticates the distribution of this species in India.

Key to the species of Sanguineum Alliance

1. Dwarf undershrub up to 0.3 m high; leaves obovate or obovate-spathulate, 1 - 3 x 0.6 - 1.8 cm, glabrous on both the surfaces; calyx 1 - 3 mm, glabrous, not cup-shaped; corolla pale-yellow; ovary glabrous .... *trilectorum*

1a. Small shrub up to 2 m high; leaves elliptic, 3.5 - 9 x 1 - 3 cm, upper surface glabrous, lower surface with a thin brownish or whitish indumentum; calyx 2 - 7 mm, sparsely tomentose, cup-shaped; corolla pink to crimson-rose; ovary glandular .... *eudoxum* var. *eudoxum*

R. eudoxum Balf.f. & Forrest, Notes R.B.G. Edinb.11:62 (1919) var. **eudoxum.** Chamberlain, Notes R.B.G. Edinb. 39(2) : 401(1982). Cox & Cox, The Encyclopedia of Rhododendron Species 119 (1997).

Syn.: *R. trichomiscum* Balf.f. & Forrest, Notes R.B.G. Edinb. 12:169 (1920).

R. trichophlebium Balf.f. & Forrest, Notes R.B.G. Edinb. 13:62 (1920).

Low twiggy shrub, 0.3 - 2 m high. Branchlets tomentose and sparsely bristly-glandular. Bark greenish to brownish. Leaves 3.5 - 7 (- 8.5) x 1 - 3 cm, elliptic, thin, apex rounded, apiculate, base rounded, upper surface glabrous, lower surface with a thin, brownish, patchy indumentum; petioles tomentose. Inflorescence 2 - 6-flowered, lax; pedicels tomentose, with a few glandular-setae. Calyx 2 - 7 mm, sparsely tomentose or glandular. Corolla tubular-campanulate or campanulate, 3 - 4 cm long, rose-pink to magenta coloured, rarely white. Ovary predominantly glandular. Capsule *ca* 2 cm long.

Distribution and habitat: India : N.E. Arunachal Pradesh. China (S.E. Xizang, N.W. Yunnan). **Map 22.**

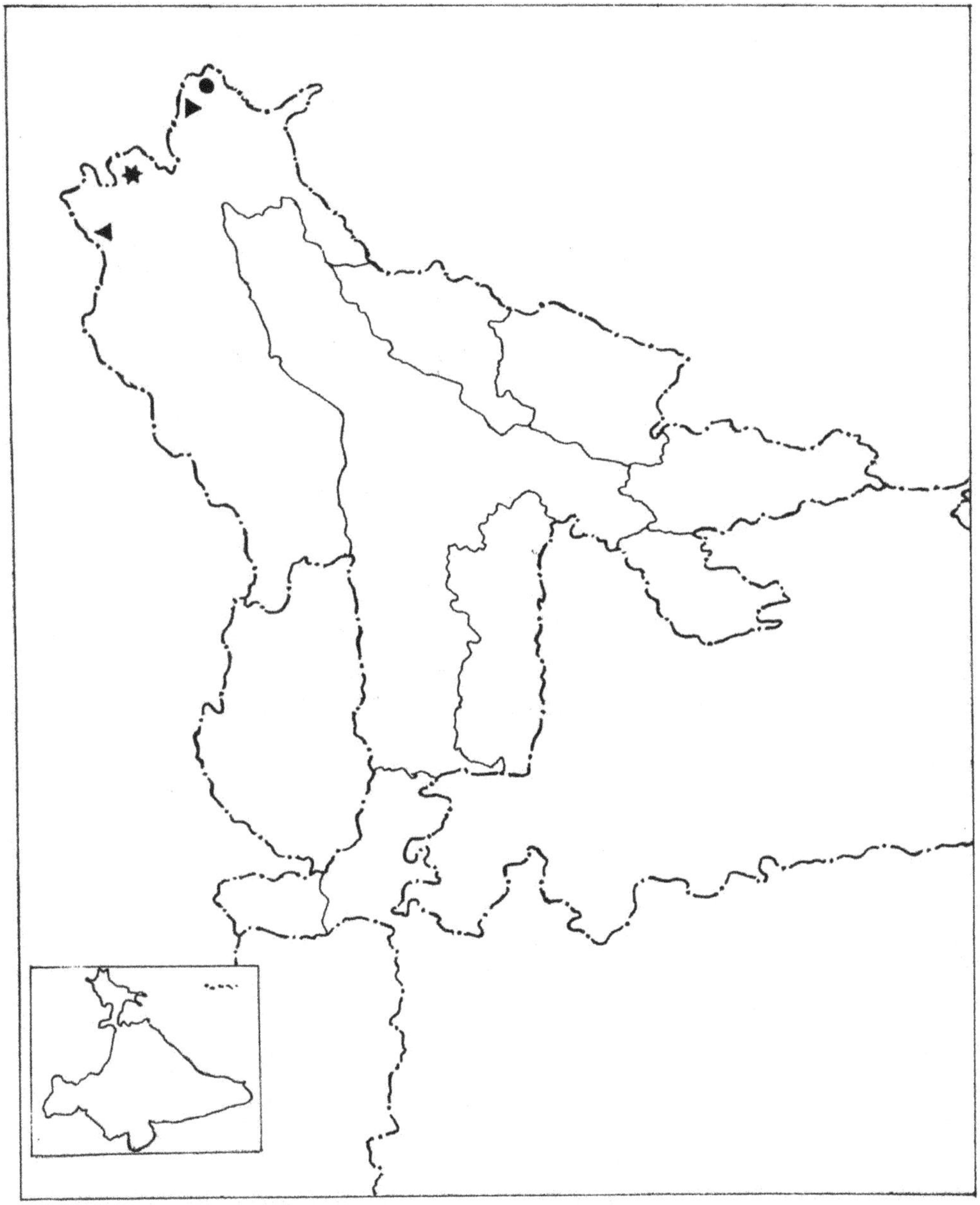

Map 22. ▼ R.beenianum. ▲ R.piercii. ● R.pocophorum var.pocophorum. ★ R.eudoxum var.eudoxum.

(Maps are only indicative and not to scale)

Notes: The species is included on the basis of the list published by Mao, *et al* in N.P. Singh & D.K. Singh (ed.), *Floristic diversity and conservation strategies in India* Vol. IV. *Angiosperms (selected groups) and Ethnobotany,* 2176 (2001). Botanical Survey of India, Calcutta.

No specimens of this species have been seen by the senior author (A.R.K.S) in the Herbaria at Kew and Edinburgh. Chamberlain & Cox do not mention its distribution in India, nor there seems to be any collections by Kingdon-Ward made during his expeditions in Mishmee Hills and other parts of N.E. India bordering S.E. Tibet, N.E. Burma and Yunnan.

R. trilectorum Cowan, Notes R.B.G. Edinb. 21:144 (1953). Chamberlain, Notes R.B.G. Edinb. 39(2) : 401(1982). Cox & Cox, The Encyclopedia of Rhododendron Species 121(1997).

Dwarf shrub, up to *ca* 0.3 m high. Leaf bud scales persistent. Leaves 1 - 3.2 x 0.6 - 1.8 cm, obovate or obovate-spathulate, apex rounded and mucronate,base attenuate, glabrous on both the surfaces; petioles winged. Inflorescence 2 - 3 (- 5)-flowered ; pedicels 2 - 3 cm glabrous. Calyx lobes fleshy, glabrous. Corolla campanulate, 3-4 cm long, pale yellow. Ovary glabrous.

Distribution and habitat: India : N.E. Arunachal Pradesh. China (S. E. Xizang). **Map 23.**

Notes: This species is included on the basis of the list of Rhododendrons in India, kindly provided by Mr. P.A. Cox to the senior author (A.R.K.S) during his visit to Glendoick.

Key to the species of Forrestii Alliance

1. Dwarf mound-forming shrub, up to 15 cm high, lateral branches creeping; leaves orbicular to obovate; inflorescence usually solitary, sometimes 2-flowered; corolla scarlet to crimson .... *forrestii* subsp. *forrestii*

1a. Upright, low shrub, up to 1.5 m high; leaves broadly ovate to elliptic; inflorescence 2 - 6-flowered; corolla pink .... *chamaethomsonii* var. *chamaethauma*

R. chamaethomsonii (Tagg & Forest) Cowan & Davidian, Rhododendron Yearbook 6:70 (1951) var. **chamaethauma** (Tagg) Cowan & Davidian, Rhododendron Yearbook 6:71 (1951). Chamberlain, Notes R.B.G. Edinb. 39(2) : 407(1982). Cox & Cox, The Encyclopedia of Rhododendron Species 126 (1997).

Dwarf shrub, 0.1 - 1 m high, compact and dense. Branchlets glandular. Leaf bud scales usually deciduous or sometimes persistent. Leaves 2 - 9 x 1.8 - 4 cm, broadly obovate to elliptic, apex rounded or retuse, base rounded to broadly cuneate, surfaces glabrous;

petioles glandular. Inflorescence (1-)2 - 5-flowered; pedicels 1 - 2 cm. Calyx minute, 1 mm or less. Corolla campanulate, 2.5 - 4.5 cm, pale pink to deep pink. Ovary densely hairy.

Distribution and habitat: India : N.E. Arunachal Pradesh. China (S.E. Xizang & N.W. Yunnan). Along steep hill slopes, rocky slopes in between 3,400 - 4,600 m. **Map 23.**

Notes: The species name *chamaethomsonii* means dwarf *thomsonii,* referring to its dwarf habit.

R. forrestii Balf.f. ex Diels, Notes R.B.G. Edinb. 5:211 (1912) subsp. **forrestii**

Dwarf mound forming, creeping shrub, 15 cm high. Lateral stems up to 60 cm long. Leaf-bud scales persistent. Leaves 1 - 2.8 x 0.9 - 1.8 cm, ovate to orbicular, apex rounded to retuse, sometimes mucronate, base broadly cuneate, upper surface glabrous, lower surface green or crimson-purple, glabrous or with a few stipitate glands; petioles 0.5 - 0.8 mm long, glandular or sparsely tomentose. Inflorescence single-flowered, or rarely 2-flowered; pedicels 10 - 20 mm, stipitate-glandular. Calyx minute, 1mm. Corolla fleshy, tubular-campanulate, 3 - 3.5 cm long, scarlet to crimson. Ovary densely glandular and rufous-tomentose. Capsule 15 - 20 x 6 - 7 mm.

Distribution and habitat: India : N.E. Arunachal Pradesh. China (S.E. Xizang, N.E. Yunnan), N.E. Upper Burma. On steep rocky slopes, moist stony pastures and on boulders, in the altitudinal range of 3,000 - 4,600 m. **Map 23.**

Notes: The species is named after the celebrated Scottish botanist-explorer George Forrest of the Royal Botanical Garden, Edinburgh who made extensive rhododendron collections in China, Upper Burma and adjoining regions.

Neriiflorum Alliance

R. neriiflorum Franchet, Bull. Soc. Bot. France 33:230 (1886) subsp. **phaedropum** (Balf.f. & Farrer) Tagg in Stevenson (ed.) The Species of Rhododendron 533 (1930). Chamberlain, Notes R.B.G. Edinb. 39(2) : 410(1982). Cox & Cox, The Encyclopedia of Rhododendron Species 130 (1997). **(Photo No. 71).**

Syn.: *R. phaedropum* Balf.f. & Farrer, Notes R.B.G. Edinb. 13:283 (1922). *R. tawangense* Sahni & Naithani, Indian Forester 105:685 (1979)

Shrub or small tree, 1 - 6 m high. Branchlets sparsely floccose-tomentose, eglandular or rarely setose-glandular. Leaves 8 - 10 cm long, oblanceolate, upper surface glabrous, lower surface usually glabrous, glaucous-papillate; petioles glandular or not. Inflorescence 5 - 12-flowered; pedicels 10 - 15 mm, sparsely stellate-tomentose, with a few glands. Calyx sparsely glandular, distinct, coloured. Corolla tubular-campanulate, fleshy, 3.5 - 4.5 cm long, straw-yellow, tawny-orange to scarlet. Ovary tomentose and with a few glands; style glabrous. Capsule 20 - 25 x 4 mm, strongly curved.

Distribution and habitat: India : Arunachal Pradesh. Central Bhutan, Upper Burma, China (S. Xizang & N.W. Yunnan). In mixed and conifer forests, amongst rocks and scrub, in the altitudinal range of 2,000 - 3,400 m. **Map 23.**

Notes: *R. neriiflorum* meaning with 'flowers like those of Oleanders', and its subspecies are widely grown in British gardens for the very attractive foliage and elegantly coloured flowers. The species is a parent of some of the choicest clones favoured in cultivation.

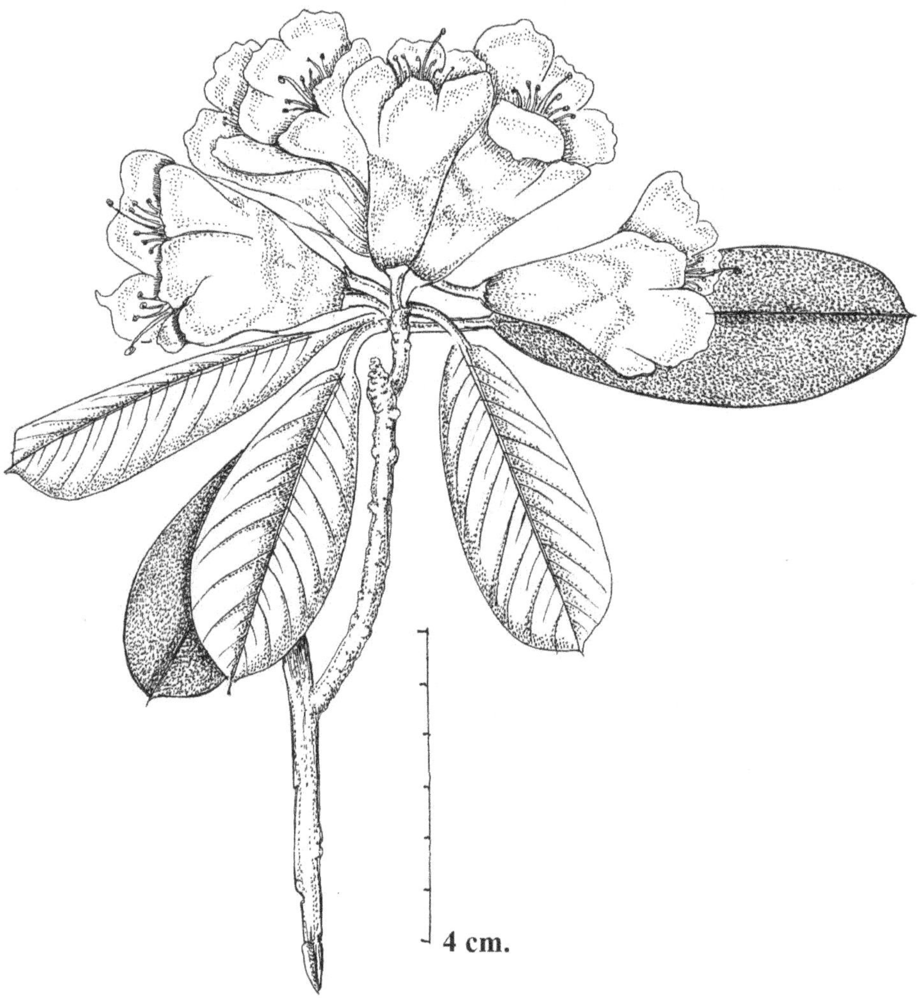

R.neriiflorum Franch.

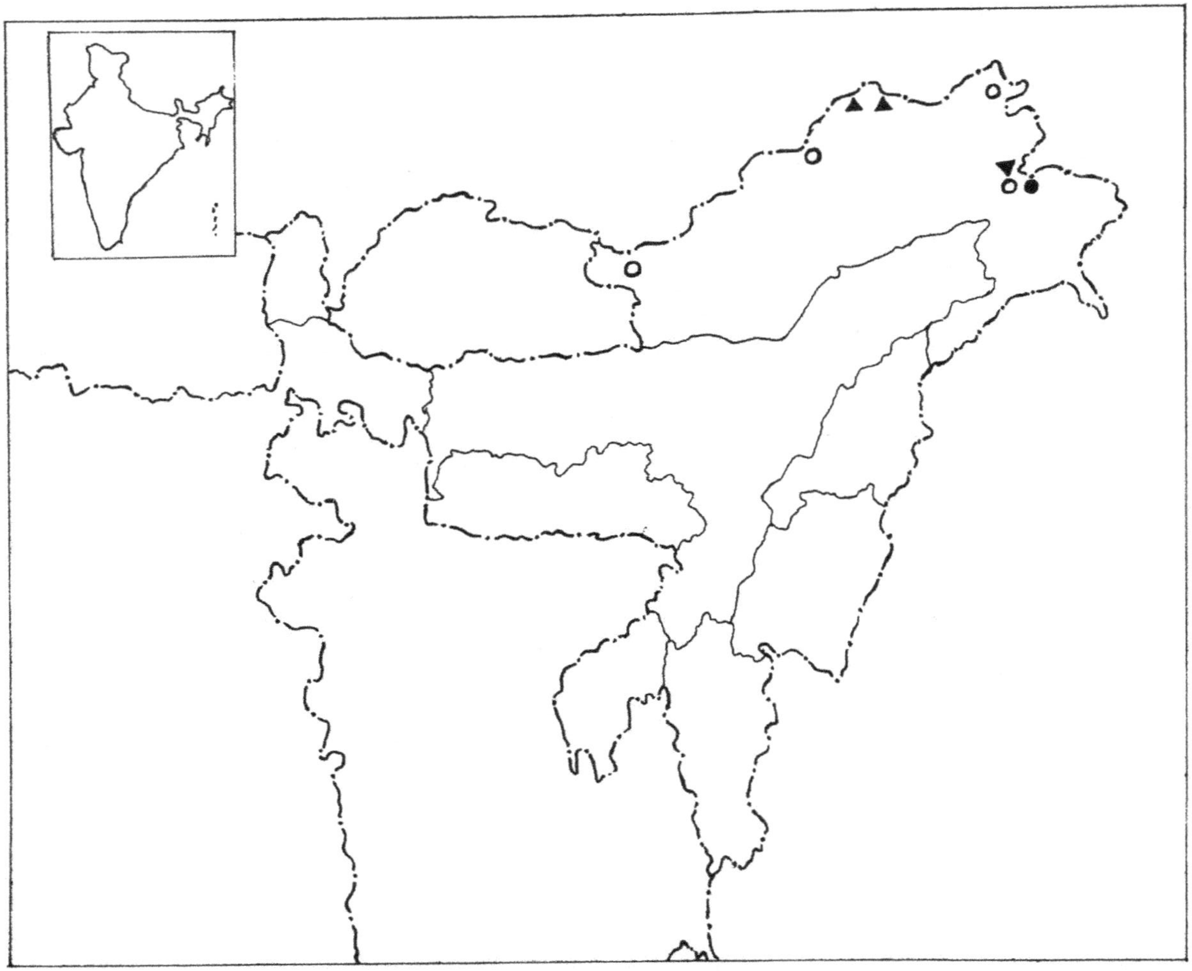

Map 23. ▲ R.trilectorum. ● R.chamaethomsonii var.chamaethauma. ▼ R.forrestii ssp.forrestii. O R.neriiflorum ssp.phaedropum.

(Maps are only indicative and not to scale)

Subsection Parishia

Subsection **Parishia** Sleumer, Bot. Jahrb. 74:548 (1949).

Syn.: Series *Irroratum* subseries *Parishii* sensu Tagg in Stevenson (ed.) The Species of Rhododendron 331, 361(1930). Chamberlain, Notes R.B.G. Edinb. 39(2) : 375(1982). Cox & Cox, The Encyclopedia of Rhododendron Species 133 (1997).

Shrubs or small trees, 2 - 10 m high. Young shoots stellate-tomentose, sometimes setose-glandular. Leaves elliptic to broadly-obovate or lanceolate, upper surface glabrous at maturity, lower surface glabrescent or with a thin stellate tomentum sometimes intermixed with glands, especially along the midrib. Inflorescence (5-) 10 - 15 (-20)-flowered. Calyx usually small, 1 - 5 mm, glandular, sometimes sparsely hairy. Corolla funnel-campanulate, up to 5.5 cm long and wide, 5-lobed with prominent nectar pouches, red to deep scarlet with deeper spots; pedicel glabrous or glandular. Stamens 10. Ovary densely tomentose and glandular; style tomentose and glandular.

Type species : *R. parishii* C.B. Clarke

This subsection is represented in India only by *R. elliottii* Watt

R. elliottii Watt in Brandis, Indian Trees 410 (1906). Chamberlain, Notes R.B.G. Edinb. 39(2) : 379(1982). Cox & Cox, The Encyclopedia of Rhododendron Species 133 (1997).

Straggling shrub or small tree, 3 - 9 m high. Young shoots stellate-tomentose and stipitate-glandular. Leaves 8.5 - 11 x 2.5 - 6 cm, lanceolate, oblong-elliptic to elliptic, apex rounded and apiculate, base rounded,upper surface glabrous,lower surface glabrous or with vestiges of stellate hairs especially near the midrib; petiole *ca* 2.5 cm long, floccose and glandular, becoming glabrous on maturity. Inflorescence 10 - 20-flowered, dense; pedicels *ca* 1cm long, stipitate-glandular. Calyx 3 - 4 mm, glandular, lobes rounded, glandular-ciliate. Corolla funnel-campanulate, with 5 prominent nectar pouches, 4 - 5.5 cm long and wide, scarlet to crimson (purplish-red to deep rose purple) with darker spots. Ovary densely rufous stellate-tomentose, intermixed with stipitate glands; style tomentose and glandular. Capsule *ca* 15 x 6 mm.

Distribution and habitat: India : Nagaland. Endemic and endangered due to widespread destruction of its habitats for shifting cultivation; in mixed forests between 2,400 - 2,700 m altitude **Map 24.**

Notes: This species is named after Mr. Elliott, a friend of Sir George Watt (the first Imperial Economic Botanist in India).

Subsection Taliensia

Subsection **Taliensia** Sleumer, Bot. Jahrb. 74:550 (1949). Chamberlain, Notes R.B.G. Edinb. 39(2) : 333(1982). Cox & Cox, The Encyclopedia of Rhododendron Species 155(1997).

Syn.: Series *Lacteum* Tagg in Stevenson (ed.) The Species of Rhododendron 370 (1930) & Series *Taliense* sensu Tagg, op.cit. 628. Subsection *Lactea* Sleumer, loc.cit. (1949).

Shrubs to small trees. Bark usually rough, flaky, brownish. Young shoots glabrous to densely tomentose, sometimes also with stipitate glands. Leaves linear to broadly ovate or broadly elliptic, upper surface usually glabrous on maturity, smooth or rarely bullate, lower surface with a woolly, felted or compacted indumentum composed of radiate, ramiform or fasciculate hairs. Inflorescence 5 - 20-flowered, usually dense. Calyx 0.5 - 12 mm. Corolla 5 - 7-lobed, campanulate to funnel-campanulate, (sometimes mortor-shaped as in *R. wightii*), nectar pouches absent, white to pink to purplish or occasionally yellow, often with conspicuous spots, sometimes with a purple blotch. Stamens 10 (- 14). Ovary glabrous to densely rufous-tomentose and / or glandular, style glabrous or partly glandular.

Type species : *R. taliense* Franchet

Notes: Amongst several species included in this Subsection, *R. wightii* is reported from India. Also, *R. bhutanense* D.G. Long & Bowers Lyon described as a new species in 1989 from Bhutan is reported to occur in Arunachal Pradesh and included here on the authority of Mr. Peter A. Cox (pers. com. to A.R.K.S). The two species, namely, *R. wightii* and *R. bhutanense* are recognized from one another as follows :

Key to the species of subsection Taliensia

1. Large shrub or small tree, 2 - 6 m high; leaves 5 - 18 x 5 - 7 cm, usually flat, base cuneate to rounded, upper surface glabrous; corolla white, cream to lemon yellow; ovary densely red-brown tomentose .... *wightii*

1a. Small shrub, 0.6 - 3 m high; leaves 6 - 12.5 x 3 - 5 cm, base cordate, upper surface with some loose hairs at the base of the midrib; corolla medium to pale pink; ovary glabrous .... *bhutanense*

R. bhutanense D.G. Long & Bowes Lyon

Cox & Cox, The Encyclopedia of Rhododendron Species 162 (1997).

Small shrub, 0.3 - 3 m high, compactly growing. Leaves 6 - 12.5 x 3 - 5 cm, broadly elliptic to obovate, with cordate base, upper surface glabrous for most part except for some loose hairs at the base of the midrib, lower surface with a slightly granular felted greyish-brown to orange-brown indumentum. Inflorescence 10 - 15-flowered; pedicels 0.8 - 1.3 cm long. Calyx minute, glabrous. Corolla campanulate, 3 - 3.5 cm long, deep pink in bud condition becoming medium to pale pink on opening. Stamens glabrous. Ovary glabrous.

Distribution and habitat: India : Western Arunachal Pradesh adjoining Bhutan. Bhutan, China (S.E. Xizang). In the altitudinal range of 3,800 - 4,300 m among scrub and other *Rhododendron* bushes in the high altitudes. **Map 24.**

R. wightii Hook.f., Rhododendrons of Sikkim Himalaya t. 27 (1851). C.B. Clarke in Hook.f., Fl. Brit. India 3:467 (1882). Chamberlain, Notes R.B.G. Edinb. 39(2) : 365(1982). Cox & Cox, The Encyclopedia of Rhododendron Species 187(1997). **(Photo No. 72).**

Shrub or a small tree, 2 - 6 m high. Branchlets thick. Bark greenish-brown to brown. Leaf and flower buds sticky. Leaves 5 - 18 x 5 - 7 cm, broadly elliptic to obovate, usually flat, apex acpiculate, base cuneate to rounded, upper surface glabrous, dark green, lower surface with a dense unistrate rust-brown compacted indumentum consisting of ramiform hairs; petioles 1 - 2.5 cm, sparsely tomentose or glabrescent. Inflorescence 10 - 20-flowered; pedicels 1.5 - 3 cm, tomentose. Calyx minute, glabrous. Corolla campanulate to mortor-shaped, usually 5-lobed, rarely up to 7-lobed, 3 - 4.5 cm long, white, cream to lemon-yellow, with light brown or purple flecks. Ovary densely red-brown tomentose; style glabrous. Capsule 2 - 3 cm long, *ca* 5 mm in diam., straight or curved.

Distribution and habitat: India : Sikkim and Western Arunachal Pradesh. Nepal, Bhutan, China (S. Xizang). In scrub vegetation, on stony ground and along ridges, in between 3,400 - 4,500 m altitude **Map 24.**

Notes: This species is named after Robert Wight, former Superintendent of the Madras Botanic Garden, well known for his publication : *Icones Indica Plantae Orientalis* or *the Drawings of Plants of East India(Wt. Ic.).*

Subsection Thomsonia

Subsection **Thomsonia** Sleumer, Bot. Jahrb. 74:545 (1949). Chamberlain, Notes R.B.G. Edinb. 39(2) : 416(1982). Cox & Cox, The Encyclopedia of Rhododendron Species 189(1997).

Syn.: Series *Thomsonii* subseries *Thomsonii* sensu Tagg in Stevenson (ed.) The Species of Rhododendron 735(1930).

Shrubs or small trees, 0.5 - 14 m high. Bark smooth or peeling. Leaves orbicular to broadly elliptic, glabrous at maturity, lower surface sometimes strongly papillate or with a thin dendroid indumentum, or with red, punctate hair bases on the lateral veins appearing as small red flecks. Inflorescence 1 - 15-flowered, usually loose. Calyx usually well-developed, up to 15 mm, cup-shaped, often coloured, lobes conspicuous. Corolla campanulate to tubular-campanulate, 5-lobed, fleshy, with 5 nectar pouches, white or yellow or pink to dark crimson, with or without flecks. Stamens 10. Ovary glabrous to glandular or rufous-tomentose; style usually glabrous, or sometimes glandular its entire length. Capsule straight.

Type species : *R. thomsonii* Hook.f.

Most of the species in the subsection are widely grown for their showy flowers and often fine smooth bark.

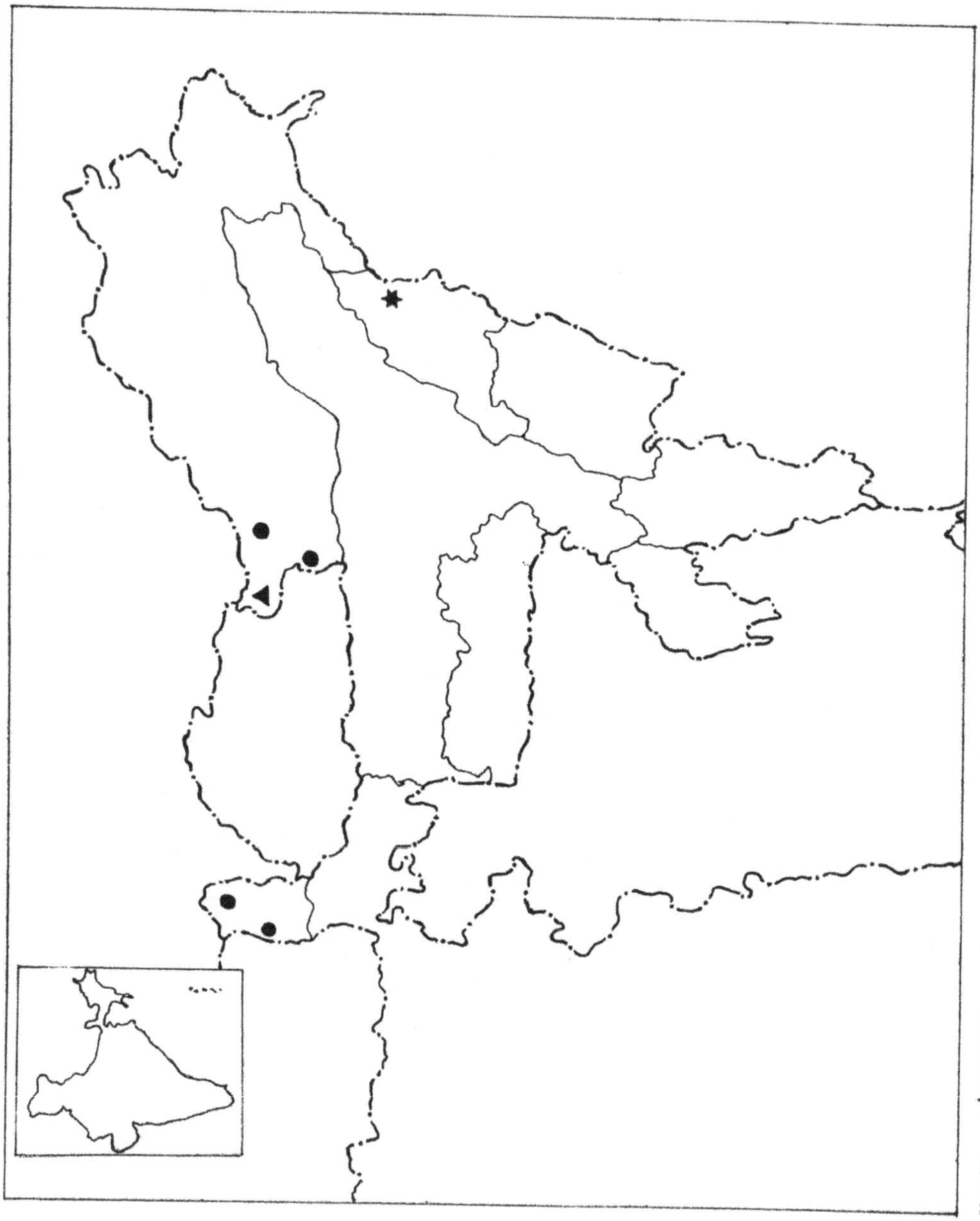

Map 24. ★ R.elliottii. ▲ R.bhutanense. ● R.wightii.

(Maps are only indicative and not to scale)

Key to the species of subsection Thomsonia

1.	Calyx glabrous	2
1a.	Calyx stipitate-glandular or ciliate	3
2.	Lower surface of leaves glaucous, papillate, glabrous	*thomsonii*
2a.	Lower surface of leaves with isolated tufts of fasciculate hairs (like hooks) on lateral veins or midrib often atleast with some straight hairs on either side	4
3.	Leaf surfaces glabrous; inflorescence 3 - 5-flowered; calyx glandular-stipitate	*cerasinum*
3a.	Leaf upper surface glabrous, lower surface with numerous unstalked glands overlying the veins with a vestige of hairs, otherwise glabrous; inflorescence 10 - 15-flowered; calyx margins ciliate	*subansiriense*
4.	Lower surface of leaves glabrous, midrib often at least with straight hairs on either side	*eclecteum*
4a.	Lower surface of leaves with tufts of fasciculate hairs	*hookerii*

R. cerasinum Tagg, Notes R.B.G. Edinb. 16:188(1931). Chamberlain, Notes R.B.G. Edinb. 39(2) : 418(1982). Cox & Cox, The Encyclopedia of Rhododendron Species 189(1997).

Shrub, 1.8 - 3.6 m high. Young shoots glabrescent. Leaves 5 - 10 x 2 - 4 cm, oblong to oblanceolate or oblong-elliptic, rounded at both ends, slightly apiculate at apex, glabrous on both the surfaces; petiole 0.7 - 2 cm long, with sparse dendroid hairs. Inflorescence lax, 3 - 5-flowered. Calyx *ca* 1.5 mm, shortly stipitate-glandular. Corolla campanulate, thick, fleshy, up to 5 cm long, crimson to scarlet or creamy white to pale pink, with reddish edges of the lobes, nectar pouches deep purple coloured. Ovary shortly stipitate-glandular; style glandular to the tip. Capsule *ca* 2 cm long.

Distribution and habitat: India : Eastern Arunachal Pradesh. China (S.E. Xizang), N.E. Upper Mynamar. In dense thickets and conifer forests in the altitudinal range of 3,000 to 3,700 m. **Map 25.**

Notes: This species is easily recognized from its bicoloured flowers and dark nectar pouches. *Cerasinum* means cherry coloured referring to its flower colour.

R. eclecteum Balf.f. & Forrest, Notes R.B.G. Edinb. 12:205(1920) var. **eclecteum**. Chamberlain, Notes R.B.G. Edinb. 39(2) : 425(1982). Cox & Cox, The Encyclopedia of Rhododendron Species 191(1997).

Syn.: *R. brachyandrum* Balf.f. & Forrest, Notes R.B.G. Edinb. 13:32(1920). *R. eclecteum* Balf.f. & Forrest var. *brachyandrum* (Balf.f. & Forrest) Cowan & Davidian, Rhododendron Yearbook 6:169(1951).

Bushy shrub, 0.6 - 2.4 m high, with coloured smooth flaking bark. Young shoots with bright red leaf bud scales. Leaves 5 - 15 x 2 - 6 cm, obovate to obovate-lanceolate or elliptic, apex rounded, apiculate or retuse, base cuneate to rounded, upper surface glabrous, lower surface usually glabrous with atleast some straight hairs on either side of the midrib, lateral veins conspicuous; petiole 0.4 - 1 cm long, narrowly winged. Inflorescence 6 - 12-flowered; pedicels 10 - 25 mm long, glabrous or sparsely glandular. Calyx 2 - 15 mm long, cupular, glabrous. Corolla tubular-campanulate, 3 - 5 cm long, white, pink, rose or purple, red or yellow, heavily spotted or flushed or stained imparting a bicoloured effect. Ovary densely stipitate-glandular; style glabrous. Capsule 15 - 25 x 6 - 10 mm.

Distribution and habitat: India : E. Arunachal Pradesh. China (N.W. & W. Yunnan, S.E. Xizang, S.W.Sichuan), N.E. Myanmar. Usually in the open on bouldery slopes, in thickets and bamboo and cane brakes, in the altitudes of 3,000 - 4,300 m.

Notes: *Eclecteum* means to be chosen out.

R. hookerii Nuttall in Hooker's J. Bot. Kew Gard. Misc. 5:359(1853). Chamberlain, Notes R.B.G. Edinb. 39(2) : 422(1982). Cox & Cox, The Encyclopedia of Rhododendron Species 193(1997).

Shrub or small tree, 3 - 6 m high. Bark smooth, grey and maroon. Leaves 6 - 15 x 3 - 7 cm, broadly oblanceolate, apiculate at apex, rounded at base, upper surface glabrous, lower surface with isolated tufts of fasciculate hairs on lateral veins; petiole 1.5 - 2 cm, glabrous, slightly winged. Inflorescence 8 - 18-flowered; pedicels 8 - 15 mm long, glabrous. Calyx 10 - 20 mm, cupular, greenish or yellowish, eglandular and glabrous. Corolla fleshy, tubular-campanulate, 3.5 - 4.5 cm long, deep rose to deep crimson, with darker nectar pouches. Ovary and style eglandular.

Photo No. 48. *R.campanulatum* ssp.*campanulatum.*

Photo No. 49. *R.campanulatum* ssp.*campanulatum.*

Photo No. 50. *R.campanulatum* ssp.*campanulatum* var.*album*.

Photo No. 51. *R.campanulatum* ssp.*aeruginosum*.

Photo No. 52. *R.wallichii*.

Photo No. 53. *R.campylocarpum* ssp.*campylocarpum*.

Photo No. 54. *R.campylocarpum* ssp.*campylocarpum*.

Photo No. 55. *R.campylocarpum* ssp.*campylocarpum* – a rare white-flowered form.

Photo No. 56. *R.arizelum*.

Photo No. 57. *R.falconeri* ssp.*falconeri*.

Photo No. 58. *R.falconeri* ssp.*falconeri*.

Photo No. 59. *R.falconeri* ssp.*falconeri*.

Photo No. 60. *R.falconeri* ssp.*falconeri*. (with cream coloured flowers)

Photo No. 61. *R.hodgsonii*.

Photo No. 62. *R.hodgsonii.*

Photo No. 63. *R.decipiens.*

Photo No. 64. *R.decipiens.*

Photo No. 65. *R.griffithianum.*

Photo No. 66. *R. griffithianum.*

Photo No. 67. *R.fulgens.*

Photo No. 68. *R.grande.*

Photo No. 69. *R.kendrickii.*

Photo No. 70. *R.lanatum.*

Photo No. 70a. *R.lanatum.*

Photo No. 71. *R.neriiflorum* ssp.*phaedropum*.

Photo No. 72. *R.wightii*.

Photo No. 73. *R.subansiriense.*

Photo No. 74. Type specimen of *R.subansiriense* (Photographed at Edinb. Herb.)

Photo No. 75. *R.thomsonii* ssp.*thomsonii*.

Photo No. 76. *R.thomsonii*.

Photo No. 77. *R.thomsonii* ssp.*thomsonii* – habit.

Photo No. 78. *R.thomsonii* ssp.*thomsonii* – a red-flowered form.

Distribution and habitat: India : W. Arunachal Pradesh. Along sheltered forest slopes with *Abies* and other rhododendrons in the altitudes between 3,000 - 3,700 m. **Map 25.**

Notes: The species is closely related to *R. subansiriense* but can easily be distinguished by its tufts of hairs or hooks on the under surface of leaf and the eglandular calyx and ovary. It is named in honour of Sir J.D. Hooker, one of the most distinguished botanist-traveller and a former Director of the Kew Botanical Garden.

R. subansiriense Chamberlain & Cox, Notes R.B.G. Edinb. 36(1) : 124(1978). Cox & Cox, The Encyclopedia of Rhododendron Species 197(1997). **(Photo Nos. 73, 74).**

Small tree, to 14 m high. Bark smooth or peeling, grey to red. Leaves 7-13 x 2-4 cm, oblong, apex rounded, shortly apiculate, base nearly rounded, upper surface glabrous, lower surface epapillate with numerous red punctate hair bases (unstalked glands) over lying the veins, each with the vestige of a branched hair; petiole 1 - 1.5 cm, glabrous. Inflorescence 10 - 15-flowered, fairly compact, pedicel 7 - 10 mm, glabrous. Calyx 4 - 5 mm, cupular, lobes rounded, margins hairy. Corolla tubular-campanulate, *ca* 4 cm long, thick with nectar pouches, crimson-scarlet, with a few purple spots. Ovary densely tomentose, eglandular; style glabrous. Capsule not known.

Distribution and habitat: India : Subansiri District, on the ridge, S.E. of Ziro (Apathani valley), Arunachal Pradesh. In temperate, mossy, rain forest with other rhododendrons and *Magnolia campbellii* in the altitudes of 2,500 - 2,750 m. Endemic. **Map 25.**

Notes: This species was collected by the senior author (ARKS) during the joint botanical exploration in Subansiri district in April,1965 with the Coxes. It is characteristically devoid of any mossy growth and with very fine light red smooth trunk with thinly peeling bark.

R. thomsonii Hook.f., Rhododendrons of Sikkim Himalaya t. 12(1851) subsp. **thomsonii.** Chamberlain, Notes R.B.G. Edinb. 39(2) : 419(1982). Cox & Cox, The Encyclopedia of Rhododendron Species 198(1997). **(Photo Nos. 75-78).**

Shrub or small tree, 1.5 - 7 m high. Bark smooth, peeling, usually of mixed colours. Leaves 5 - 11 cm long, glabrous thick and leathery, orbicular, ovate to broadly elliptic, apex rounded, minutely apiculate, base rounded to cordate, upper surface dark to medium green, glabrous, lower surface glaucous-papillate, glabrous; petiole glabrous or sparsely glandular. Inflorescence 3 - 10-flowered, dense; pedicels 10 - 25 mm, glabrous or glandular. Calyx large, cup-shaped, 10 - 20 mm, variously coloured. Corolla fleshy, campanulate, deep blood-red or dark wine-red coloured, with 5 nectar pouches at the base, 3.5 - 6 cm long, more or less spotted and with white or white-flushed pink bloom. Ovary glabrous or glandular; style glabrous. Capsule 1.5 - 2.5 cm long.

Distribution and habitat: India : Sikkim, Arunachal Pradesh. E. Nepal, Bhutan, China (S. Xizang). In Rhododendron and *Abies* forests in between 2,400 - 4,300 m altitude **Map 25.**

Notes: It is named after T. Thomson, a former Superintendent of Calcutta Botanical Garden. The subsp. *thomsonii* freely hybridizes in the wild with several other species in the Himalayan region, especially in Sikkim and Bhutan. *R. x candelabrum* Hook.f. is a hybrid between subsp. *thomsonii* and *R. campylocarpum* subsp. *campylocarpum.* Similarly *R. sikkimense* Pradhan and Lachungpa, found in North Sikkim is considered to be a natural hybrid between subsp. *thomsonii* and *R. arboreum,* exhibiting the characters of both the parent species. It resembles *R.thomsonii* in its cup-shaped calyx, blood-red coloured flowers; it comes closer to *R. arboreum* in its oblong-ovate to elliptic, leathery leaves with a thin layer of matty, silvery to yellowish-brown indumentum on the lower surface of leaves. In *R. thomsonii* the leaves are orbicular or broadly ovate to obovate with glabrous upper surfaces and glaucous under surfaces.

Another species viz., *R. blumei* Nuttall, from Arunachal Pradesh is imperfectly known as the Type specimen is stated to be sterile (Chamberlain, op. cit.)

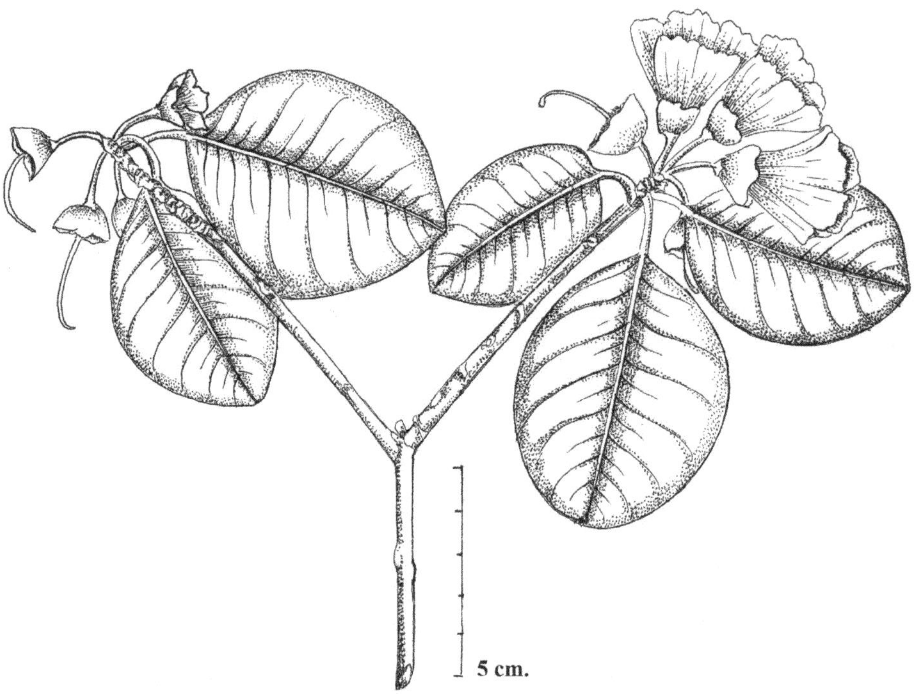

R. thomsonii Hook.f.

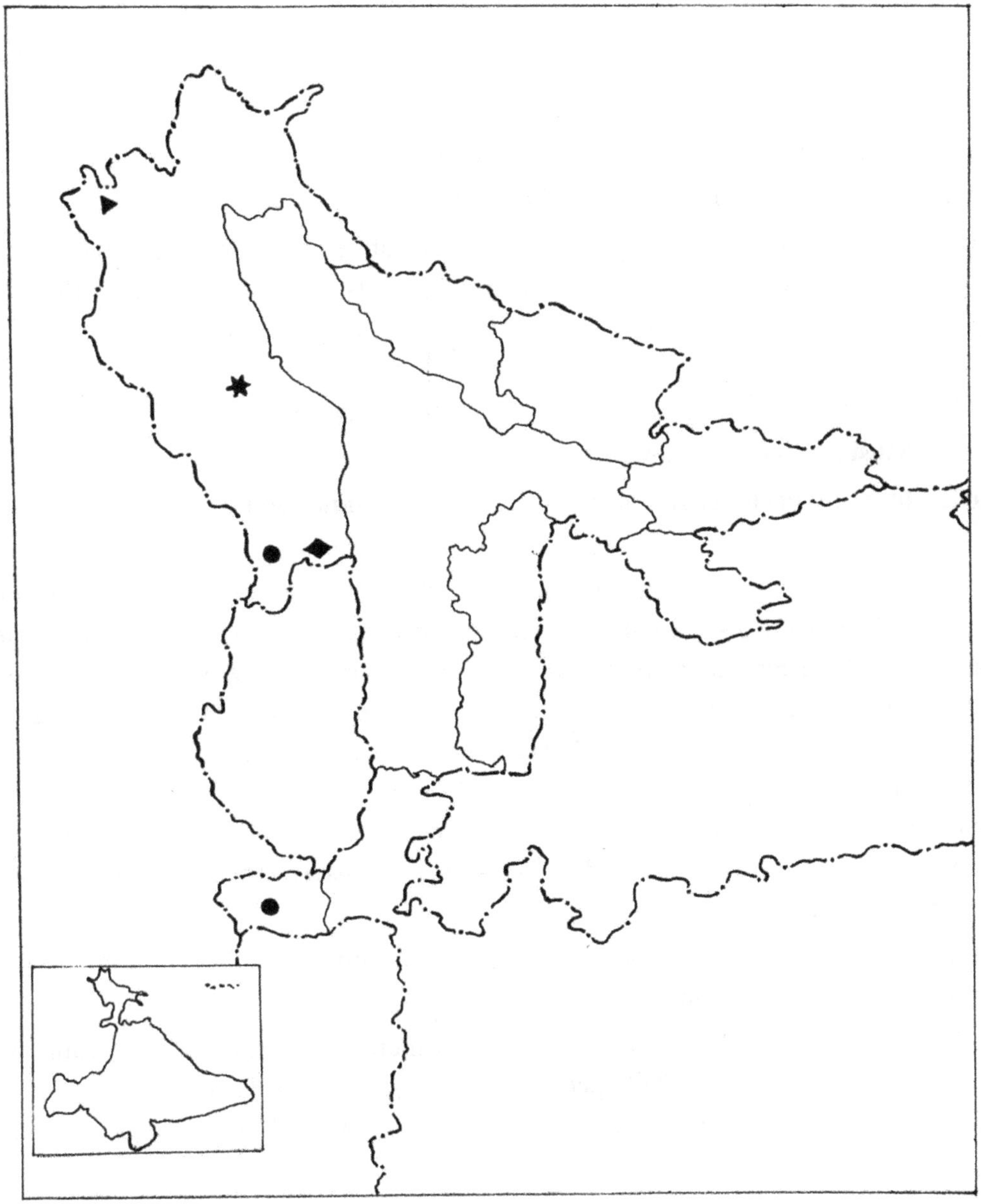

Map 25. ▼ R.cerasinum. ◆ R.hookerii. ✱ R.subansiriense. ● R.thomsonii ssp.thomsonii.

(Maps are only indicative and not to scale)

6
Subgenus Tsutsusi

Subgenus **Tsutsusi** (Sweet) Pojarkova in Fl.URSS18:55 (1952).Cox&Cox,The Encyclopedia of Rhododendron Species 364(1997).

Shrubs, creeping or erect, 0.5 - 4.5 cm high. Vegetative buds and inflorescence from within the same bud scales. Leaves persistent or deciduous, alternate, covered with either simple or ribbon-like flattened bristles or stiff, glandular hairs. Inflorescence 1 - 15-flowered. Corolla rotate to tubular-campanulate. Stamens (4-) 5 - 10 (- 12). Ovary bristly to glandular hairy, style usually glabrous. Seeds without wings or tails.

Section **Tsutsusi** (Sweet) Pojarkova

Rhododendron arunachalense Chamberlain & Rae, Edinb. Journ. Bot. 47(2) : 131, 1990.

Shrub, 4 - 5 m high. Young shoots adpressed-brown-strigose. Leaves probably dimorphic, 3.5 - 6.5 x 1.6 - 2.8 cm, chartaceous, spring leaves elliptic-ovate lanceolate, apex acuminate, base cuneate, margins with stiff incurved hairs, more or less entire, upper surface sparsely adpressed-strigose, lower surface adpressed rufous-strigose. Inflorescence 3-flowered; pedicels 5 - 7 mm, densely brown-strigose with large stiff and thinner crisped hairs. Calyx obsolete, strigose. Corolla open funnel-shaped, light pink, darker inside, 2.3 to 2.7 cm, tube 6-7 mm, glabrous, 3 mm broad at base, 6 mm broad above. Stamens 5, 3 cm long, scabrid at base. Ovary densely strigose; style glabrous, 3.5 cm. Capsule unknown.

Distribution and habitat: India : Subansiri district, Yachuli, Arunachal Pradesh. In river beds, 1500 m altitude. Endemic. **Map 26.**

Notes: Probably allied to *R. saxicolum* but with a more acute leaf, darker corolla and glabrous corolla tube within. This species is geographically isolated and is the most westerly of the subgenus and section *Tsutsusi* which is confined to N. Myanmar, W.,S. and E. China to Taiwan and Japan.

7
Subgenus Azaleastrum

Subgenus **Azaleastrum** Planchon, Fl. De Serrus. 9:75(1853/4). W.R. & M.N. Philipson, Notes R.B.G. Edinb. 44(1) : 2(1986). Cox & Cox, The Encyclopedia of Rhododendron Species 1(1997).

Section **Choniastrum** Franchet, Bull. Soc. Bot. Fr. 3:229(1886).

Type species: *R. stamineum* Franchet

R. moulmainense Hook.f. is the only species in India representing the section and the subgenus *Azaleastrum*.

R. moulmainense Hook.f. Bot. Mag. 82:t.4904(1856).

Syn.: *R. stenalum* Balf.f. & Forrest, Notes R.B.G. Edinb. 10:157(1917).

Shrub or small treee, 15 m high. Young branches glabrous. Leaves 5 - 15 x 2 - 5 cm, oblong-lanceolate or nearly elliptic, shortly acuminate, cuneate at base, glabrous, margins of adult leaves rarely sparsely hirsute; petiole 1 – 2 cm long. Inflorescence axillary, (1-) 3 - 8-flowered, at the ends of leafy shoots. Pedicel glabrous, 1.5 - 2 cm long. Calyx minute, glabrous, 5-lobed. Corolla funnel-shaped with a narrow tube, 3.5 - 6 cm long, 5-lobed, lobes broad, spreading, 2 - 4 cm long, white, with a pink to violet tinge, with or without yellow or green blotch, scented. Stamens 10, filaments hairy below. Ovary 0.5 cm long, glabrous; style glabrous. Capsule narrowly cylindric, *ca* 7 cm long.

Distribution and habitat: India : Arunachal Pradesh (Subansiri District). China (S.E. Xizang, Yunnan), Myanmar, Taiwan, Japan, S. Indo-China to W. Malaysia. In evergreen and subtropical conifer-mixed forests in the altitudinal range of 50 - 3750 m. **Map 26.**

Notes: A widely distributed and originally recorded from Moulmain, S. Myanmar, this species was first collected in profuse flowering condition by one of the authors (A.R.K.S.) from the hill ranges surrounding the Apathani valley in Subansiri district in April, 1965 at an altitude of approximately 1600 m and was reported as a new record for India under the name *R. stenaulum.* Only one or two trees were found to be growing in *Pinus wallichiana*-Bamboo-mixed broad leaved forests along the hill slopes.

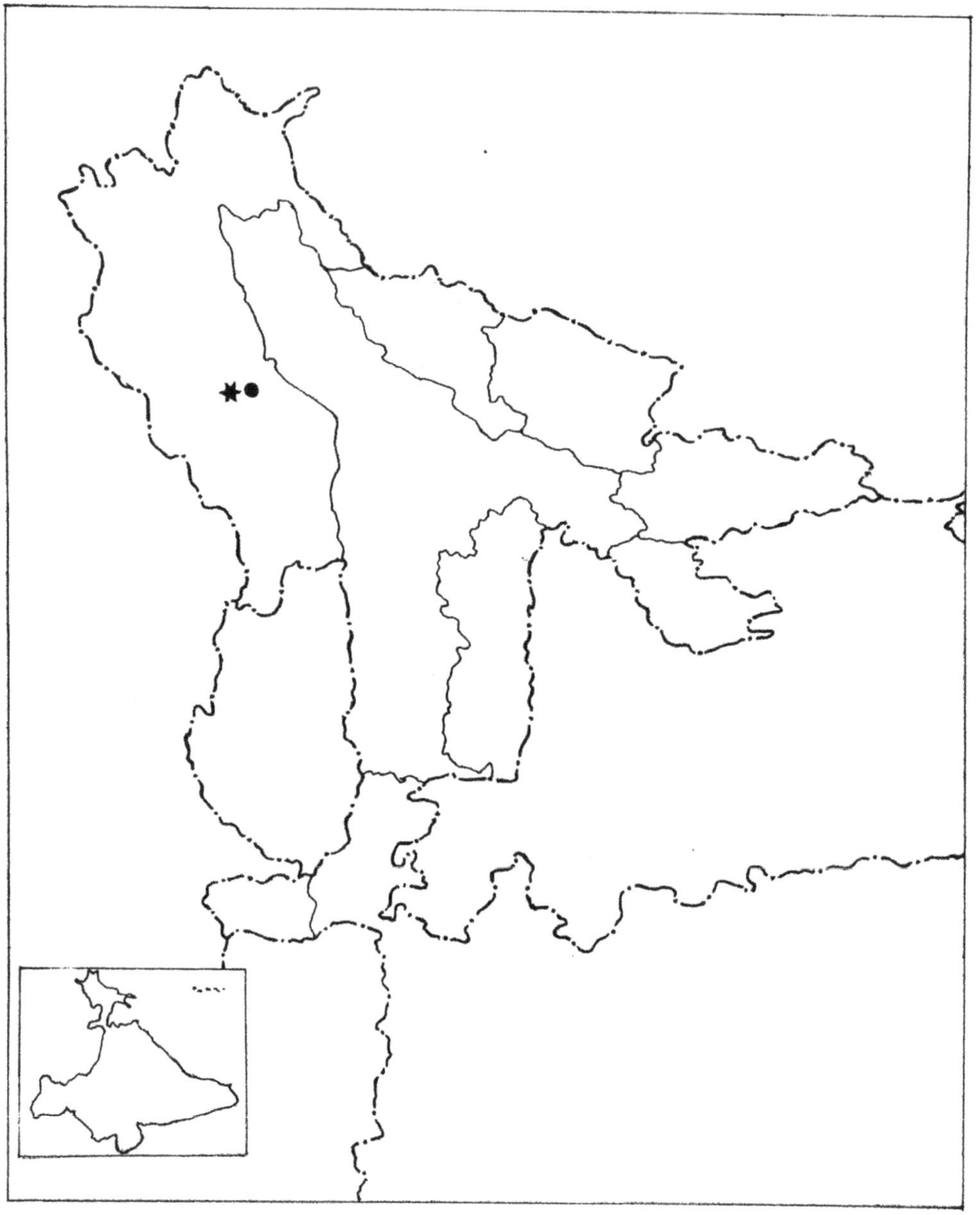

Map 26. ★ R.arunachalense. ● R.moulmainense.

(Maps are only indicative and not to scale)

Photographs of some colour drawings from J.D. Hooker's Rhododendrons of Sikkim Himalaya (1849)

Plate No. 1. *R. camelliflorum.*

Plate No. 2. *R. cinnabarinum.*

Plate No. 3. *R. cinnabarinum* ssp. *cinnabarinum* (*R. roylei*).

Plate No. 4. *R. edgeworthii.*

Plate No. 5. *R. pendulum.*

Plate No. 6. *R. lepidotum* (*R. salignum, R. elaeagnoides*).

Plate No. 7. *R. ciliatum.*

Plate No. 8. *R. pumilum.*

Plate No. 9. *R. virgatum.*

Plate No. 10. *R.arboreum* ssp. *cinnamomeum* var. *cinnamomeum* (*R. campbelliae*).

Plate No. 11. *R. barbatum.*

Plate No. 12. *R. wallichii.*

Plate No. 13. *R. campylocarpum.*

Plate No. 14. *R. falconeri.*

Plate No. 15. *R. fulgnes.*

Plate No. 16. *R. wightii.*

Plate No. 17. *R. Thomsonii.*

Plate No. 18. *R. x candelabrum.*

Photographs of some Type materials in Kew Herbarium

R. vaccinioides (Herb. Kew)

R.edgeworthii (Herb. Kew)

R.dahousiae (Herb. Kew)

R.exasperatum (Herb. Kew)

R.barbatum (Herb. Kew)

R.kendrickii (Herb. Kew)

R.subansiriense (Herb. Kew)

R.pendulum (Herb. Kew)

R.beenianum (Herb. Kew)

R.thomsoni ssp. thomsoni (Herb. Kew)

Glossary of Botanical Terms used

Accrescent	:	increasing in size with age
Acicular	:	needle-shaped
Actinomorphic	:	symmetrical in several planes
Acuminate	:	tapering into a long slender point, usually of a leaf
Acute	:	with a pointed tip
Adpressed	:	(hairs) held close to a leaf or stem and parallel to its surface, but not adherent to it
Agglutinate	:	(indumentum) compacted and looking as if painted or glued on
Apiculate	:	(leaf) with a sharp, short pointed apex
Arachnoid	:	(indumentum) with matted hairs appearing like a cob-web
Bistrate	:	(indumentum) in two-layers usually the lower layer is thin, plastered and the upper layer is thicker.
Bract	:	a modified, reduced leaf on an inflorescence, often bearing a flower in its axil
Bullate	:	(leaf) having a blistered or puckered appearance as in the case of upper surface of the leaf of *R. edgeworthii*
Calyx	:	the outer most series of flower parts
Campanulate	:	bell-shaped
Capitellate	:	(hairs) mop-headed, which make up the indumentum
Capsule	:	a dry dehiscent fruit enclosing seeds
Caudate	:	with a tail-like tip
Chartaceous	:	papery

Cilia	:	marginal hairs
Ciliate	:	with hairy margins
Contiguous	:	touching one another (as in the case of scales)
Cordate	:	(leaf) heart-shaped i.e. with a deep notch at the base
Cordulate	:	slightly cordate
Coriaceous	:	(leaf) leathery, thick
Corolla	:	the tube and petal lobes of the flower
Corymb	:	a type of inflorescence with flower-stalks arising at different levels but attain the same height to form a flat-topped domed cluster
Crenulate	:	(scales) with small teeth or notches
Cuneate	:	(leaf base) wedge-shaped
Cuspidate	:	tipped with a sharp, rigid point
Deciduous	:	falling off; not persistent
Declinate	:	(style) bent or curved downwards or forward
Deflexed	:	pointing downwards; the style of species in subsection *Lepidota*
Dehiscent	:	(fruit) splitting into definite parts, usually length-wise
Dendroid	:	tree-like, branching (referring to hairs)
Deciduous	:	falling off, not persistent
Dimorphic	:	occurring in two forms, usually in the case of leaves
Disciform	:	disc-shaped, usually describing the calyx
Divaricate	:	very much divergent
Elepidote	:	without scales, opposite of lepidote; a distinctive taxonomic character

Elliptic	:	(leaf) about two times longer than broad tapering equally at both the ends
Epiphyte	:	a plant merely growing on another plant
Erose :		eroded
Exerted	:	projecting beyond, usually referring to the stamens and / or style projecting from the corolla
Fasciculate	:	(hairs) clustered or growing in bundles
Felted	:	(indumentum) matted with intertwined hairs
Ferrugineous	:	rust coloured
Filiform	:	thread-like
Fimbriate	:	(hairs) fringed with slender processes or appendages
Floccose	:	clothed with patchy, loose tufts of hairs or wool, referring to leaf indumentum or tomentum on young shoots and leaves
Fusiform	:	spindle-shaped
Glabrescent	:	with deciduous hairs becoming glabrous
Glabrous	:	smooth; without hairs or glands
Glandular	:	hairs with a small round sticky knob
Glaucous	:	(leaf) green, strongly tinged with bluish grey or greyish waxy bloom; sea-green or blue-green
Granular	:	consisting of small grains
Heteromorphic	:	not uniform in kind / structure
Hirsute	:	bearing long, distinct hairs
Hypocrateriform	:	salver-shaped
Impressed	:	somewhat sunken, describes the way the style joins the ovary abruptly at the base of the slender style as opposed to tapering gradually into it ; an important charater in the identification of certain species

Indumentum	:	a covering, such as of hairs
Inflorescence	:	a cluster of flowers
Lacerate	:	teary
Lanate	:	clothed with woolly and intergrown hairs
Lanceolate	:	shaped like a lance-head (usually of leaves)
Lepidote	:	clothed with small scurfy scales; a very significant character in the classification of rhododendrons
Ligulate	:	strap-shaped
Loriform	:	protective covering, usually of hairs of varied types
Monotypic	:	represented by a single category such as section, subsection or species etc. in the classification
Mucronate	:	(leaf) abruptly terminating into a sharp point
Nectaries	:	sac-like vessels in the corolla base containinig a sweet honey-like secretion, as in most species of subsection *Thomsonia*
Oblanceolate	:	(leaf) widest near the tip gradually tapering to base ; reverse of lanceolate shape
Oblique	:	with unequal sides
Oblong	:	(leaf) longer than broad with the sides more or less parallel and with rounded ends
Obovate	:	reverse of ovate
Obsolete	:	scarcely present; often used to describe the calyx
Obtuse	:	blunt, usually referring to the tip of a leaf, calyx lobe etc.
Ovary	:	the ovule-bearing part of the flower later develops into the fruit with seeds
Ovate	:	(leaf) egg-shaped with stalk attached to the wider and with the tip pointed

Papillose	:	with nipple-like superficial projections on the leaf surface
Pedicel	:	the stalk of a flower in an inflorescence
Perulae	:	Leaf bud scales which enclose the terminal bud in winter, which fall off when new leaves appear
Petiole	:	leaf-stalk
Pilose	:	covered with fine soft hairs, as in some leaves
Prostrate	:	lying along the ground; procumbent
Pruinose	:	having a bloom or waxy powdery secretion on the surface
Puberulous	:	slightly hairy with short hairs
Pubescent	:	clothed with short soft hairs
Punctuate	:	marked wit dots, depressions or glands
Raceme	:	inflorescence with flowers evenly spaced on rachis, with the flower stalks (pedicels) of equal length with the flowers opening in succession from base to the apex
Rachis	:	the main stalk of the inflorescence
Radiate	:	(hairs) spreading from a common centre
Ramiform	:	(hairs) branched
Reticulate	:	like a network
Reticulum	:	a net-like structure
Revolute	:	(leaf) rolled back from the margin or the tip
Rosulate	:	clustered, collected into a rosette
Rotate	:	wheel-shaped
Rugulose	:	with small wrinkles
Scabrid	:	warty or covered with small hard points

Scales	:	found on the leaves of lepidote rhododendrons; important taxonomic character
Sessile	:	without stalk (usually of leaves & flowers)
Setae	:	bristles
Setose	:	bristly
Spathulate	:	narrowly oblong with broader end; like a spatula
Stamen	:	the floral organ bearing the anther and pollen
Stellate	:	(hairs) star-shaped
Stigma	:	the terminal part of style which receives the pollen
Stipitate	:	(glands) elevated on a stalk
Stoloniferous	:	producing suckers from the basal branches which creep along the surface of the ground, taking roots at intervals, giving rise to new shoots ; a means of vegetative propagation in some species
Subcoriaceous	:	(leaf) somewhat leathery
Suborbicular	:	nearly rounded
Taxon	:	(plural taxa) a general term for a taxonomic group of any rank eg. genus, species, subspecies, variety sharing certain distinctive characters
Terete	:	cylindrical
Tomentose	:	densely matted with tomentum
Tomentum	:	pubescence composed of matted woolly hairs
Truncate	:	as though cut off at the end
Undulate	:	wavy-margined, usually of leaf, petal or calyx lobes
Unistrate	:	single-layered (as in indumentum)
Velutinous	:	with fine hairs; velvety
Ventricose	:	(corolla) inflated or swollen on one side
Vesicular	:	bladder-shaped, referring to scales

Villous	:	shaggy with long, weak hairs
Viscid	:	sticky
Zygomorphic	:	symmetrical on either side of a median plane only

Selected References

Anon.(1963). The Rhododendron Year Book. R.H.S.,London.

Chamberlain David, Roger Hyam, George Argent, Gillan Fairweather & Kerry S. Walter (1996). *The Genus Rhododendron : Its classification & synonymy.* Royal Bot. Gard., Edinburgh.

Chamberlain, D.F. & S.J. Rae (1990). A revision of Rhododendron IV. Subgenus Tsutsusi. *Edinb. Journ. Bot.* 47(2) : 89 – 200.

Chamberlain, D.F. (1982). A Revision of Rhododendron II. Subgenus Hymenanthes. *Notes R.B.G. Edinb.* 39(2) : 209 - 486.

Cowan, J.M. (1942). The Rhododendrons of India. *150 Ann. Vol. Roy. Bot. Gard. Cal.* Parts 1 & 2, Calcutta.

Cox, P.A. & Kenneth, N.E. Cox (1997). *The Encyclopedia of Rhododendron Species.* Glendoick Publishing.

Cox, P.A. (1985). *The Smaller Rhododendrons*. Batsford.

Cox, P.A. (1990). *The Larger Rhododendrons.* Batsford.

Cullen, J. & D.F. Chamberlain (1978). A Preliminary Synopsis of the Genus Rhododendron. *Notes R.B.G. Edinb.* 36 : 105 – 126.

Cullen, J. & D.F. Chamberlain (1979). A Preliminary Synopsis of the Genus Rhododendron. *Notes R.B.G. Edinb.* 37 (2) : 327 – 338.

Cullen, J. (1980). Revision of Rhododendron I. Subgenus Rhododendron, sections Rhododendron and Pogonanthum. *Notes R.B.G. Edinb.* 39(1) : 1 – 207.

Davidian, H.H. (1982). *The Rhododendron Species Vol. I, Lepidotes*. Batsford.

Davidian, H.H. (1989). *The Rhododendron Species Vol. II, Elepidotes*, Batsford.

Davidian, H.H. (1992). *The Rhododendron Species Vol. III, Elepidotes contd.* Batsford.

Davidian, H.H. (1995). *The Rhododendron Species Vol. IV, Azaleas.* Timber Press.

Hooker, J.D. (1849). *The Rhododendrons of Sikkim-Himalaya.* London.

Hooker, J.D. (1854). *Himalayan Journals,* Vols. 1&2. London.

Janaki Ammal, E.K., I.C. Enoch & M. Bridgewater (1950). Chromosome numbers in species of Rhododendron. *The Rhododendron Year Book* 5 : 78 – 91. R.H.S., London.

Kingdon-Ward, F. (1935). Rhododendron seeds with special reference to their classification. *Journ. Bot.* 73 : 241 – 247.

Kingdon-Ward, F. (1947). Observations on the classification of the genus Rhododendron. *The Rhododendron Year Book.* R.H.S., London.

Leppik, Elmer E. (1974). Evolutionary interactions between Rhododendrons, pollinating insects and rust fungi. *Quarterly Bull. Amer. Rhodo. Soc.* 28 : 70 – 89.

Mao,A.A.,K.P.Singh & P.K.Hajra(2001). Rhododendrons.In:N.P.Singh & D.K.Singh(Ed.). *Floristic Diversity and Conservation Strategies in India, vol. IV. Angiosperms (selected groups) and Ethnobotany.* Bot.Surv. India, Calcutta.

Magor, E.W.M. (1986). Sikkim Himalayan Rhododendrons in British Gardens. *Himal. Plant Journ.* 4(11 – 12) : 58 – 68.

Philipson, M.N. & W.R. Philipson (1973). A history of Rhododendron classification. *Notes R.B.G. Edinb.* 32(2) : 223 – 238.

Philipson, M.N. & W.R. Philipson (1982). A preliminary synopsis of the genus Rhododendron. *Notes R.B.G. Edinb.* 40 : 225 – 227.

Philipson, M.N. & W.R. Philipson (1986). A revision of Rhododendron III. Subgenera Azaleastrum, Mumeazalea, Candidastrum and Therorhodion. *Notes R.B.G. Edinb.* 44(1) : 1 – 23.

Poston, C. (1996)(Ed.) *The Rhododendron Story.* R.H.S., London.

Pradhan, U.C. & Sonam T. Lachungpa (1990). *Sikkim-Himalayan Rhododendrons.* Primulaceae Books, Kalimpong.

Rao, R.S. & J. Joseph (1965). Observations on the Flora of Siang Frontier Division, N.E.F.A. *Bull. Bot. Surv. India* 7 (1 – 4) : 138 – 161.

Sastry, A.R.K., S.K. Kataki (1967). On some species of Rhododendron from Subansiri district, N.E.F.A. Indian For. 93(4) : 264 – 265.

Sastry, A.R.K., S.K. Kataki; P.A. & P. Cox, P. Hutchison (1968). Rhododendron santapaui sp. nov. from subansiri district, N.E.F.A., India. Journ. Bombay Nat. Hist. Soc. 65 (3) : 742 – 747.

Sastry, A.R.K. & P.K. Hajra (1983). Rare and endemic species of Rhododendron in India. In : S.K. Jain & R.R. Rao (Ed.) *An assessment of threatened plants of India.* Bot. Surv. India, Calcutta.

Sleumer, H. (1968). In : *Flora Malaysiana ser.* 1, 6 : 474 – 668.

Stevenson, J.B.(1930)(Ed.). *The Species of Rhododendron.* The Rhododendron Society, London.

Names in roman bold type refer to currently accepted taxa. Those in italics are synonyms.

V

W

X

www.ingramcontent.com/pod-product-compliance
Lightning Source LLC
Chambersburg PA
CBHW080929240926
51849CB00003BA/416
* 9 7 8 9 3 5 2 3 0 0 4 2 6 *